SSADM VERSION 4: A User's Guide—Second Edition

THE McGRAW-HILL INTERNATIONAL SERIES IN SOFTWARE ENGINEERING

Consulting Editor

Professor D. Ince
The Open University

Titles in this Series

SSADM Version 4:
A User's Guide—Second Edition

Malcolm Eva

McGRAW-HILL BOOK COMPANY

London · New York · St Louis · San Francisco · Auckland · Bogotá
Caracas · Lisbon · Madrid · Mexico · Milan · Montreal · New Delhi
Panama · Paris · San Juan · São Paulo · Singapore · Sydney
Tokyo · Toronto

Published by
McGRAW-HILL Book Company Europe
Shoppenhangers Road · Maidenhead · Berkshire · SL6 2QL England
Tel 0628 23432   Fax 0628 770224

British Library Cataloguing in Publication Data

Eva, Malcolm
   SSADM Version 4: User's Guide.—2Rev.ed.
   —(McGraw-Hill International Series in Software Engineering)
   I. Title   II. Series
   658.40380285421
   ISBN 0–07–707959–0

Library of Congress Cataloging-in-Publication Data

Eva, Malcolm
   SSADM version 4: a user's guide / Malcolm Eva. – 2nd ed.
     p.   cm. – (The McGraw-Hill international series in software engineering)
   Includes bibliographical references and index.
   ISBN 0–07–707409–2
   1. System design.   2. System analysis.   I. Title.   II. Series.
   QA76.9.S88E9   1994
   004.2′1–dc20   93–46575 CIP

12345   CL   97654

Typeset by Computape (Pickering) Ltd, North Yorkshire
and printed and bound in Great Britain by Clays Ltd, St Ives plc

Contents

Preface to first edition

Scope

Since this book's conception, its scope has changed and narrowed to the form it appears in today. Originally, it was to be a detailed exposition on version 4 of SSADM, descriptions of how it could and should be implemented in organizations, and finally a detailed discussion of its strengths and weaknesses, comparing it with other systems analysis methods.

When version 4 was launched in June 1990, it was clear that the changes were so extensive that the book could not keep its intended scope and stay a commercial proposition. So, I have cut ambitious plans to produce *the* standard volume on methodology and confined myself to a description of SSADM version 4. I did this reluctantly at first, but by the time I was half-way through the draft, I felt only relief that I had not let myself in for the rest as well.

My first feelings about version 4 were ambivalent: I had not expected it to change as much from version 3 as it has, and was furious that much cunningly preprepared work for this book, based on version 3, all had to be scrapped. There was a lot that was new; that was daunting. Some things that I knew well had changed, or lost importance; that made me a little indignant on their behalf. New complexity seemed to have been built in for no good reason, and I thought that it would be a far harder method to sell to organizations.

Further examination of the material changed my views—not in one bound, but gradually and surely. The weaknesses in version 3 had been addressed: Dialogue Design and Physical Design were two such areas. Both have been changed completely and given a much better chance of achieving their objectives. The analyst is not locked into views of the current system, as was the case with version 3; rather, the emphasis is on identifying and meeting the requirements for the new system.

Any project undertaken using SSADM could not now put a half-trained team of analysts on to a job, with virtually no Project Management support and expect it to deliver the goods, as I have seen happen with several methods, including SSADM version 3. When such projects failed, SSADM was usually blamed. This was unfair, always, because the team were not allowed to use SSADM techniques properly, were not given reasonable estimating guidelines, and received no clear leadership. Version 4 has met that particular dragon and disposed of it with its new structure and mandatory interfaces with Project Management.

The deeper I looked into the new face of SSADM the more I saw the strength of the new approach. The increased rigour in the definition of all SSADM products and cross-checks led to a more reliable end-product, while at the same time the modular structure and the stress on non-SSADM activities led to greater flexibility, both in the scope of projects undertaken and in the emphasis each separate installation might place on particular aspects, such as ergonomics, or the human–computer interface.

The packaging of the enhanced method embraces a comprehensive manual, a SSADM dictionary and two entity–relationship–attribute meta-models of the method. This book does not attempt to outdo that packaging; its purpose is to condense and clarify the new structure of SSADM, and, using a continuous worked example, demonstrate the techniques employed.

I have not gone into the history of SSADM in any detail; neither have I rehearsed the justification for using structured methods of analysis and design. I have started from the assumption that a reader understands already that a method for systems analysis and design is preferable to the traditional seat-of-the-pants approach. What I expect a reader to ask is not 'Why should I use a structured method? but 'What is SSADM version 4?' That is the question I hope this book will satisfy.

Structure of the book

The book is in three parts: Part 1 introduces SSADM and describes the new Structural Model. Each of the five modules that make up the core method has a chapter of its own that describes the inputs, outputs, stages, steps, and tasks employed.

Part 2 of the book describes the techniques. Each technique has a chapter to itself. A possible exception occurs in the description of Entity/Event Modelling, Chapter 12. Two techniques are described here: Entity Life History Analysis, and Effect Correspondence Diagrams. Each is a facet of Entity/Event Analysis, but the techniques themselves are very different.

Part 3 comprises a chapter on applying SSADM within an organization, and then a glossary of SSADM terms. The glossary is more substantial than that in many books, which tend to list a dozen common words and feel their duty is done. The source for this glossary is largely—but not exclusively—the dictionary volume in the *SSADM Reference Manual*. As with any specialism, the field of SSADM is rich with jargon, some of which is familiar from other contexts, and therefore possibly misleading. I have tried to incorporate all terms that are likely to be used in the course of a SSADM project. I am sure that many have slipped through the net, even so.

Acknowledgements

It is impossible to write a book of this nature without the help and support of others, even though it is for the most part a solo—and gruelling—effort. I would like to express my thanks to the following people who, in one way or another, helped me to reach this stage: Andy Ware of McGraw-Hill, for his constant encouragement when I feared I was losing my way; Caroline Ashworth of AIMS and Duncan Quibell of Brighton Polytechnic for giving the manuscript a much-needed technical review, and making valuable suggestions as to accuracy and presentation; colleagues at Telecom ACT for general encouragement and specific advice, particularly Malcolm Pither and John Rogers; Diane and Natalie for patiently putting up with my vanishing from normal homelife every night for several months while I wrote this book, and supporting me all the while; Natalie for her help with the typing.

Preface to second edition

This book was written soon after the formal launch of SSADM version 4, and the publication by NCC Blackwell of the reference manual. In common with many of the projects to which I refer, there were urgent time constraints which caused a number of errors to leap into the text and diagrams, and some cursory descriptions of complex matters, which deserved more thorough explanation.

None the less, the reception of the book has on the whole been very gratifying. When I was invited to consider a second edition, I was delighted to be given a chance to put right some of the embarrassing errors (as they seemed to me), and to give some topics more justice than I had done first time round.

More importantly, since the first edition was produced, the method has been used in a number of different sites and environments, and is now a tried and proven method, rather than a theoretical one. Weaknesses and woolliness in the *SSADM Reference Manual* have had to be addressed, and new approaches considered to specific problems in, for example, Entity/Event Modelling. In this revision, I have tried to incorporate some of these new thoughts, sometimes to discuss them, and sometimes, if they are complex issues, just to draw attention to them.

This book is addressed still to new practitioners of SSADM, and students learning about this approach to systems analysis and design. I hope that older hands too will find it a useful guide both to the structure of the method and to the specific techniques.

I have kept the structure of the first edition, with the exception of Chapter 8, Tools and techniques. I have incorporated the material from this chapter on the three techniques of Data Flow Modelling, Logical Data Modelling, and Entity Life History Analysis into the relevant chapters on those techniques. Otherwise, although the contents of each chapter may have varied or grown, the structure of the book is unchanged. I have also included more exercises at the end of certain chapters, particularly on Function Definition and Dialogue Design.

Acknowledgements for second edition

The following people have, wittingly or otherwise, given me much help in the revision of this text, and I would like to express my appreciation and gratitude to them all: Phil Lomax, of Model Systems, Ron Segal of Pantechnicon, Dave King, of University of Brighton, Andy Tremain, of University of South West, and Roger James, formerly of BT Customer Systems. My thanks go too to Jackie Harbor and Jenny Ertle of McGraw-Hill Book Company for their editorial encouragement of this enterprise, and also to the production and editorial team at McGraw-Hill.

I am grateful to LBMS for their permission to use and describe the Jackson structure notation, which is their copyright, and which is central to several graphical techniques of SSADM.

PART ONE   Structure of SSADM

1. Introduction to SSADM

1.1 Aims of chapter

This chapter discusses the following topics:
- The history of SSADM
- The principles of SSADM
- The drive towards version 4 of SSADM, launched in 1990

As the thrust of this book is towards a description of SSADM in its current state, the reader will find that almost all subsequent chapters deal with the structure and techniques of SSADM version 4. I have, as stated in the Preface, narrowed the scope of the text to description, rather than evaluation or discussion of comparative methods.

Although this book may be used as a pocket manual by practitioners, I think it is important for the reader not only to see SSADM in a context, but also to understand both why it was developed and the philosophy and principles behind its shape, as well as appreciating how, as a method, it has been enhanced to keep pace with the changes in the world.

1.2 Background of SSADM

History

Structured Systems Analysis and Design Method (SSADM) has been used in government computing since its launch in 1981. It was commissioned by the Central Computing and Telecommunications Agency (CCTA) in a bid to standardize the many and varied IT projects being developed across government departments. The CCTA investigated a number of approaches, before accepting a tender from LBMS to develop a method based on their own product, LSDM.

Since its first appearance and adoption, SSADM has undergone changes, partly to remedy perceived weaknesses in the method, but mostly as a response to developing technology and changes in the business world. In 1986, version 3 was launched, its main change being the introduction of a Dialogue Design technique to address the growth in on-line processing, a development that earlier versions had not tackled.

Open standard

Although SSADM was designed for the government, the CCTA were keen on its becoming an open standard, freely available to other interests, in industry or in the academic community, for instance. A benefit from this is that industry then has an interest in producing support for the method. This support is principally in terms of training and automated tools. This has been mutually beneficial, to the method and to industry, and SSADM is now used widely throughout industry, as well as government.

SSADM now

SSADM version 4 was launched in 1990, since when it has been taken up by most user organizations. There are still some version 3 sites, especially those who started projects using version 3 and have elected not to migrate. In most cases now, SSADM means version 4. The Design Authority Board (DAB) is now responsible for the method's further development.

The SSADM user group plays a very important role in exploring new avenues for SSADM's evolution and growth, particularly two subgroups: future development, and the technical committee. At the time of writing, areas that are being explored by the user group for future enhancement are object-orientation, distribution, expert systems and soft systems.

1.3   SSADM philosophy

Throughout SSADM's history, certain features have been central to it. These have been carried forward through all versions, and are still at the heart of version 4. New emphases have been added, but they supplement rather than supersede the central features. These constant features can be summed up under three headings:

- Three views
- Avoid lock-in
- User involvement

These headings have been well rehearsed since 1981, with the method's first public appearance, so to maintain continuity, I shall define briefly what each means. We shall see later that each has, in fact, been enhanced by a combination of greater rigour and pragmatism.

Three views

The views in question are views of the system data. They represent function, structure, and the effects of time, as shown in Fig. 1.1.

Various philosophies of systems analysis emphasize a particular approach to the system: structured analysis, as propounded by Yourdon, De Marco and others, explores the functionality of a system and, using techniques such as Data Flow Diagrams, investigates the transformation of system data as a result of inputs from outside the system boundary.

This can be seen as a form of information plumbing, with pipelines (or data flows) carrying the data from sources to sinks, via stations that carry out some sort of transformation on that data.

Another view, the data analysis view, says it is all well and good looking at the functionality of systems, but the important thing is the underlying structure of the data. To understand that, we have to build an entity model (or, in SSADM terms, a *Logical Data Structure*). Once the data structure is in place, it will stay constant, regardless of what else happens in the business. Procedures and functions may change, after all, but the data must always be there to support the business activities, whatever they may be.

The third view says that yes, it is fine recognizing data structures, and seeing how the plumbing carries data from one part of the system to another, as and when it is needed, but both models are static, giving snapshots of possibilities. What is really

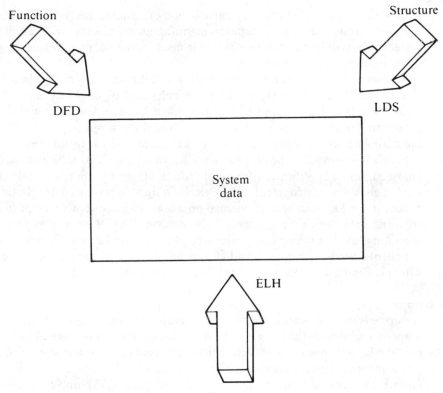

Figure 1.1

needed is a way to model business and integrity rules as they affect the data. Accordingly, we must model the effects of time on our data, and show how each item can be brought into being, how it can be amended, and how it can be removed from the system, according to the business rules of the application area. This third view gives us the *Entity Life History* (ELH).

Each of the philosophies places one of these views above the others in importance. SSADM instead gives each an equal importance, and makes them complement each other. A series of cross-validation checks between them ensures that at the end of analysis, as rich a picture as possible is gained of the system.

The *Data Flow Diagrams* (DFDs) act as a check on the Logical Data Structure (LDS), to show that each data item is created and amended at some time. The LDS ensures that the data is present in the system, to allow the function processing shown on the DFD. The ELH subsequently identifies all of the events that impact on the system, causing changes to the state of the data. At this point, omissions from earlier investigations are frequently uncovered, and the picture expanded. Integrity rules for making amendments to data are discovered and built in.

The three views of SSADM ensure a detailed analysis of the target systems is carried out.

Avoid lock-in

It could be argued that the very early forms of systems analysis were practised by salesmen from the large computer manufacturers. Clients with an information systems (IS) requirement would find their needs being adapted to suit the prescribed solution, rather than the reverse.

Problems arising from this are obvious: the technology is driving the solution, not vice versa. Should the system need to be enhanced or modified later, the client is locked into the original supplier, and must have the extended solution dictated by the constraints of the technology once again. Should the supplier go out of business, or the hardware/software become obsolete, the client really is in trouble.

SSADM approaches the problem from the other end. The nature of the problem is analysed, and a specification for a solution is drawn up, independently of any one implementation environment. A Logical Design is produced for both data and processing, which can be implemented on a selected environment, or ported between environments, with no adjustment or alteration. The Physical Design would need amending, as that is implementation-specific, but the Logical Design, which meets the clients' needs, is constant. Enhancements are carried out on the Logical Design, which is flexible enough not to need a re-specification in the process.

User involvement

Aesop tells of a fox who was attracted by grapes growing high up a vine. In fact, the grapes were so high that however high the fox jumped, they were always just out of reach. About to give up, and declare that they were too sour anyway, the cunning fox saw a systems analyst walking down the road. As this was a tale told in Ancient Greece, the systems analyst was not yet trained in SSADM, and regarded himself as an expert on all matters on which he was consulted.

The fox explained the problem: it could not reach the grapes, which it desired to eat. 'Right you are,' said the systems analyst, 'leave it with me, and by nightfall the grapes will be in your reach. I shall have two of them as my fee.'

The fox, being cunning, asked the systems analyst if more information were needed, or could it help, but the systems analyst said, 'I can cope now, thank you very much, run along till nightfall.'

When the fox returned at nightfall, the systems analyst said that all was nearly ready, and could it pop back in a couple of hours. After two more returns, at dawn the systems analyst announced that the project was complete, and the fox could go-live and collect its grapes now.

The fox looked, and saw that the systems analyst had built a scaffolding around the vines, for the fox to climb up. 'I thought you were going to shin up yourself and pick them for me', said the fox, aggrieved. 'I can't climb up a scaffolding, foxes aren't built for that. Anyway, there isn't even a ladder to get me to the first platform. I'm not accepting this system.'

The analyst pleaded in vain that he had carried out his terms of reference, that the fox had not asked him to pick the grapes himself.

The fox was so upset that it even contemplated eating the systems analyst, but thought he would probably be too sour, so it slunk off, vowing never to ask an expert for help again.

Had the systems analyst been SSADM-trained, he would have understood that the user is the expert in the applications area. He would have ensured that the fox's requirements were thoroughly analysed and understood before he designed a solution, and would have offered the fox a set of possible solutions so that the fox itself could have chosen the way to reach the grapes. At every point in the project where a part of the investigation or specification was completed, the fox would have been asked to inspect the results and approve them, before the next stage was begun. At the end, the fox would have had its requirements met in full, and the systems analyst would have been paid.

Where this fable departs from reality is that, too frequently, the systems analyst would have been paid before the user found that the system was not what was wanted.

1.4   A product-based approach

The new approach to SSADM, as expressed in version 4, appears radically different at first sight, but on further inspection shows itself to be close to the original. What it has done is to make explicit what was implicit before and, because it was implicit, was often overlooked or omitted. The need for a clear interface for Project Management was always there, but now the 'hooks' are explicitly defined, as are the interfaces for such activities as Risk Assessment, Configuration Management, Capacity Planning, Quality Control, etc.

No IS development project should ignore these activities, but unless they are built in to a method, the temptation to take a short cut for the sake of showing a system in operation seems irresistible.

Apart from an altered structure, the approach that SSADM has taken to build these in is to change its orientation from activity-based to product-based.

What, how, when

The practitioner is now tasked with producing predefined deliverables. The end of an activity is reached when this deliverable is accepted by the project manager (or module manager, or stage manager). The new structure of SSADM tells the project team *what* is to be produced, *when* it is to be produced, and *how* it is to be produced.

WHAT

SSADM now includes a dictionary to define every product that comes from an activity. There is a hierarchy of such products, which fall under three broad headings: *technical*, *management*, and *quality*. Figures 1.2–1.6 show the hierarchy of products under those headings.

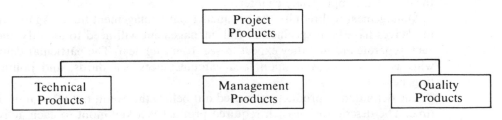

Figure 1.2

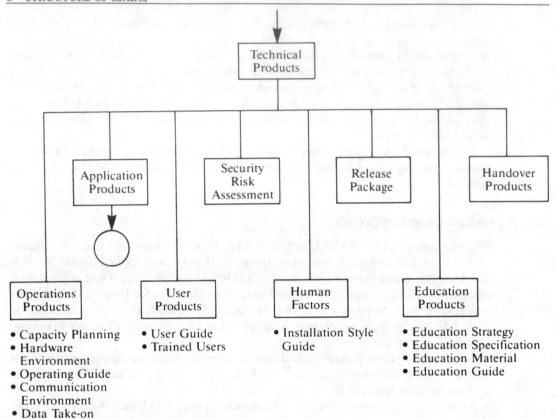

Figure 1.3

The lists are not prescriptive: local standards or practices may require additional, or alternative, deliverables. What is important is that every product which is to be delivered to the project management is to be specified under its appropriate place on the tree.

The products from the application product node are produced by the SSADM practitioners. The Product Breakdown Structure for each of the Modules is shown at the end of the appropriate chapter.

Management will require the products from management node. Again, not all of the leaves have been specified in full: management will need to identify and define each separate product they expect to see from each leaf. The particular deliverables will, as always, depend upon local circumstances, standards, and political contingencies.

The definition of products is carried out before the event, not after or at the same time. The description of each required product is a key input to each activity (see Chapter 2).

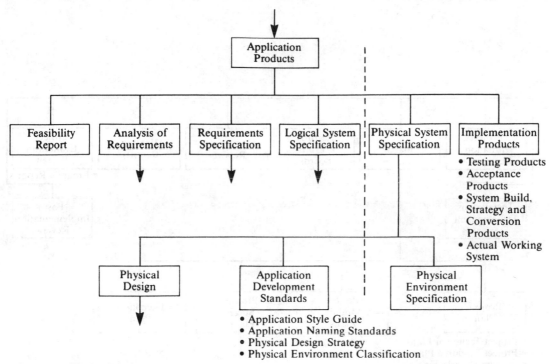

Figure 1.4

Product specification
Each product on the hierarchy will be defined in the following terms:

Title   A simple identifier, which is accessible to non-technical people as well as the technical specialists.

Purpose   A brief statement of what the purpose of the product is, e.g. to communicate the analyst's understanding to the user, or to show the layout of a screen used in a dialogue.

Composition   A list and description of all the component parts of the product. If it is a report, or form-based, a design for the form should be included, showing project header information, version numbers and release dates for Configuration Management purposes, and space for every item of information required. The *SSADM Reference Manual* (CCTA, 1990) provides suggested layouts for forms, but the actual forms' design is a matter of local standards.

Derivation   A statement of where the information in the product was derived, e.g. from discussion with users, input from Project Management or Required System LDM.

Quality   One of the key innovations to SSADM with version 4, this is the predefined criterion which the product must match if it is to be accepted. Before any activity

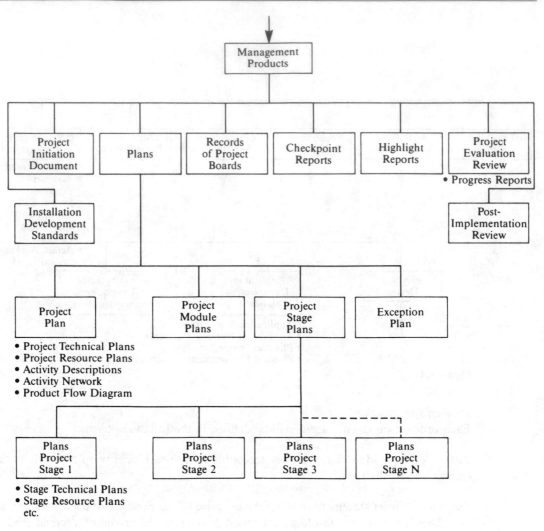

Figure 1.5

begins, the team must know what level of quality it is to meet. This might be at the level of asking if every box on a form is completed, or of checking the accuracy of data flows on a Data Flow Diagram, or of specifying the cross-checks between two techniques or products. Again, the precise criteria are tailorable for each project; the important thing is that they are set *before* the activity which produces it is started.

External dependencies   A statement of any input to the product that comes from outside core SSADM (see Chapter 2). In addition, any outside body who will need that product, such as a user at a review, may be mentioned here. Many products will have no external dependencies.

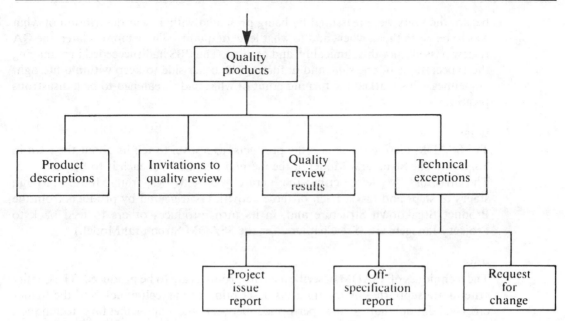

Figure 1.6

References A list of references to any subsequent SSADM activity that uses these products as input.

This emphasis on the product, and the definition of products, is both a good Project Management tool, and a good medium of gaining user support. I know from a consultancy experience in 1992 the value of a Product Breakdown Structure (PBS) in managing user expectations. This was a third-party contract, rather than an in-house development, and considerable cash and goodwill was at stake. At the end of the Requirements Analysis Module, the users of this particular project were unhappy about several aspects of the deliverables that had been passed to them for approval.

A key source of unease was the fact that they had not known what to expect, beyond some Data Flow Diagrams and a Business System Option which, to their surprise, had already been selected on their behalf. The quality review was an unhappy occasion, because of this lack of preparation. The fault lay with the project team; the user team wanted very much to cooperate, but were denied the opportunity. I was invited to help with the deadlock at this point, and to try to restore the good relations—and more important, trust—between the parties. The structure and support activities of SSADM provided the solution.

The first Module was behind the team, and deadlines did not permit a return to it. Both sides agreed readily to the project team preparing and delivering a PBS for Stage 3. This would enable the users to prepare for the QA reviews, and know what to expect; it also told the project team both what to prepare, and also to what standard to prepare it. There was no sound reason for the PBS for Stages 1 and 2 not to have been drawn up, but it was not; now, the team recognized its worth, and when Stage 3 began, the users were reassured by being presented with a clear description of what was to be given them, when, and to what level of quality. Three months later, the QA review passed smoothly, amicably, and on time. The PBS had succeeded in managing the expectations of one side, and in forcing the other side to keep within quite tight guidelines. This marked the turning point of what had threatened to be a disastrous project.

WHEN

The SSADM dictionary defines the products, as above, to tell us about the *what* in SSADM. The Structural Model of the method tells us *when* each is to be produced. The Structural Model describes the hierarchy of activities, from Module through stages to steps and tasks. Each of these activities is triggered by products from the Product Breakdown Structure and, in its turn, produces others to feed back to Project Management. (See Chapter 2 for the SSADM Structural Model.)

HOW

The techniques of SSADM describe *how* the products are to be produced. The quality criteria are supported by the rigorous description of the techniques, and the tighter cross-validation that is now performed between the complementary techniques. Automated support, such as CASE, helps too in the generation of the products, and their validation. For some of the products that support the SSADM practitioner, CCTA have produced a series of publications to illustrate the techniques concerned, and how they fit alongside the core method.

Summary

SSADM now means version 4 of the method. This version was introduced in 1990, after a series of prerelease briefings and seminars aimed at encouraging its smooth absorption into the SSADM user community.

The changes are structural and procedural, having modified the shape of the method, and adapted the techniques.

The emphasis is now on products instead of activities. The whole method can be summarized in three threads that are mutually complementary:

- Product-oriented: *what*
- Flexible structure: *when*
- Three views of model: *how*

2. Shape of SSADM

2.1 Aims of chapter

In this chapter you will learn about:
- The place of SSADM in the project life cycle
- The SSADM Structural Model
- The information highway
- Project activities
- Project Procedures and CCTA subject guides

2.2 SSADM in the project life cycle

An organization that uses SSADM in the analysis and design of its IT projects is assumed to have reached a level of maturity in its use of IT. If it is still being tentative about computerization, or, alternatively, is allowing a number of uncoordinated, *ad hoc* systems to be built, it is probably not ready to make the commitment of resources and effort needed to implement IT successfully. It may, therefore, not be ready to make the commitment necessary to practise SSADM.

Commitment to SSADM

The analysis and design stages of an IT project represent the Full Study, but by no means represent all of the project effort. If these two activities are poorly executed, the resulting system will be inefficient, ineffective and a liability. However good the programmers, if they are given inadequate specifications, they will not produce code to suit the User's requirements.

A prerequisite, therefore, for the use of SSADM is a commitment to providing the resources and organizational structure to allow SSADM to be used properly. By 'properly', I mean in such a way that it produces a proper specification of a system to meet its User's requirements, that demands minimal maintenance, that is flexible enough to incorporate later requirements, and is portable between environments. Figure 2.1 shows the nature of the project structure in such an organization.

The three levels reflect the three threads described in Chapter 1: the what, the how, and the when.

The board level, where strategy is examined and agreed defines the *what*. In other words, it describes which areas must be automated, or at least studied for change, and, in very broad terms, what the requirements are for each.

The IT directorate, at one level down, takes this *what* and makes decisions as to *how* it is to be addressed. The directorate also takes the decision about *when* each is to be tackled.

At the bottom level, the teams of service providers carry out the SSADM and

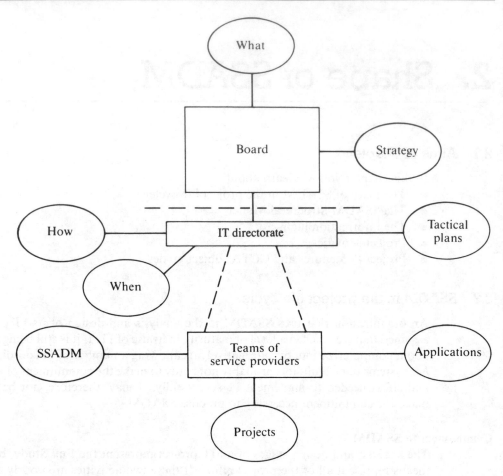

Figure 2.1

post-SSADM activities that make up the individual projects that define *what* is provided and *how* it is provided for the business requirements. The analysts, designers and programmers are all to be found here, as are other support teams that I shall mention below.

Progress of the project

The likely stages in an IS project in an organization such as described in the previous section are:

- An IS Strategy Study
- An IS Tactical Study
- A Feasibility Study
- A Full Study, to produce a specification
- A development project, that produces and tests the physical system according to the specification

Figure 2.2 shows the course of an IS project, and the place of SSADM inside it.

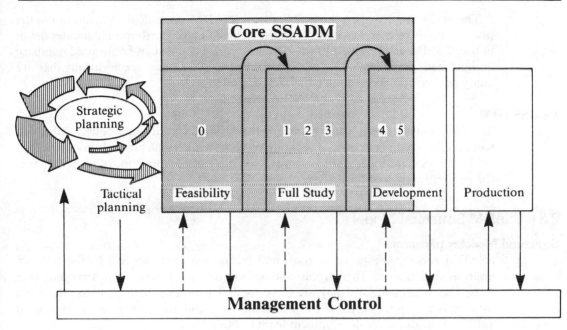

Figure 2.2

SSADM only covers the Feasibility and Full Studies. While particular SSADM techniques can be used during Strategic Study, another form of modelling the business and its broad requirements will be used then.

Feasibility

Whether or not a project is initiated as a result of an earlier Strategic Study, many will begin with Feasibility. Projects which are small and low risk are likely not to need to carry out some form of Feasibility Study to determine their course and get board commitment to them. The project team will receive their terms of reference from the findings of the study. The Feasibility team will consider the proposed project from two principal views:

- *Technical feasibility—Is the technology available to satisfy the requirement of the Users?*
- *Business feasibility—Can a sound business case be made for continuing the work? A Cost/Benefit Analysis must be made to assess the economic viability of the proposals. Assuming that the proposed project is economically feasible, an Impact Analysis must be made to decide whether or not it is operationally feasible, i.e., will it work in this environment and culture?*

If the answers to all these considerations are favourable, the Feasibility Report will direct the scope and direction of the project.

This is SSADM's first involvement with the project. Although SSADM has a Feasibility Module, with its defined stages and steps, and in this framework performs the tasks of Feasibility, there is more to Feasibility than simply the SSADM Module. The CCTA publication on Feasibility (Section 2.4) gives guidance on how SSADM is built into this study.

The next set of activities, the Full Study, is SSADM's principal contribution to the project and its success. I shall describe the shape of that contribution in greater detail in Sec. 2.3. The end products of the Full Study are a detailed and optimized database design, ready to be implemented, and a set of programming specifications that are ready to turn into code, using either a 3GL or a 4GL.

Post-SSADM

The development work, which transforms these specifications into coded and tested software, is often regarded as a separate project. While results from it may well feed back into SSADM for revision, as an activity it is outside the scope of SSADM, as is the implementation of the system.

2.3   SSADM Structural Model

Structural Model requirements

SSADM encompasses a set of tools and techniques to be applied to the tasks of analysis and design. These tools and techniques are placed in a structure that prescribes when and how they are to be used. The aim of them all is to model the information requirements of the business user, and to specify the solutions in sufficient detail for the development team to build them.

There are other techniques, apart from the standard ones for systems analysis, and the problem is how to integrate these with the SSADM techniques. The solution chosen is to have two streams of activity: the SSADM activities, in their allotted place, and a set of ancillary activities that feed in the required information to support SSADM. ·

The model

The first stream is called *core SSADM*, and it comprises a hierarchy of activities. From the top down, the hierarchy is: Module → stage → step → task. There are five Modules in this stream, ranging from Feasibility through to Physical Design. In each of these Modules there are one or more stages, with defined activities and producing defined products.

The Modules are: Feasibility Study, Requirements Analysis, Requirements Specification, Logical System Specification, Physical Design. Each is described in Chapters 3 to 7 respectively. In these chapters I shall describe in detail the stages and steps of each module, and all tasks associated with them.

Projects are essentially linear, and so the Modules are sequenced to allow the natural project lifecycle to be followed. However, each module is a self-contained set of activities, and must be managed as a discrete project. It is possible that in an IT project, each Module will go to a separate contract, so that a different supplier may be employed on each. This indicates the importance of ensuring that the end of every activity is a set of products, to the required quality, so that another supplier may then take them as a starting point to continue the project.

Each Module has its own set of plans, timescales, controls, and monitoring procedures. In addition, each Module was its own set of actors, who may not be the same through the whole project. It is highly unlikely, for example, that business

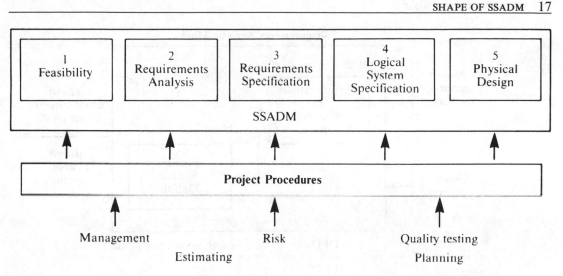

Figure 2.3

analysts who model the requirements and required system will also produce an optimized physical database design. Definition of the actors who will complete the tasks are part of the input to each Module.

The other stream is known as Project Procedures, and interfaces with core SSADM via Project Management activities. I shall say more about these procedures in Sec. 2.4. In Chapter 20 I shall describe them in more detail.

Figure 2.3 illustrates this relationship between the two streams.

Core Modules

The five Modules are defined in terms of the inputs and products. Inside each module the decomposed activities are defined, also in terms of their inputs and products. Figures 2.4–2.8 show the five Modules, with the inputs, products, and stages.

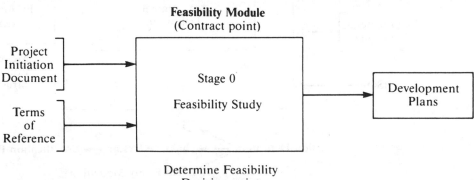

Figure 2.4

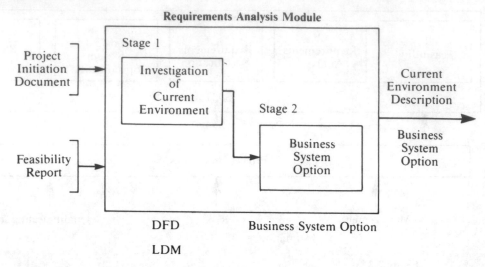

Figure 2.5

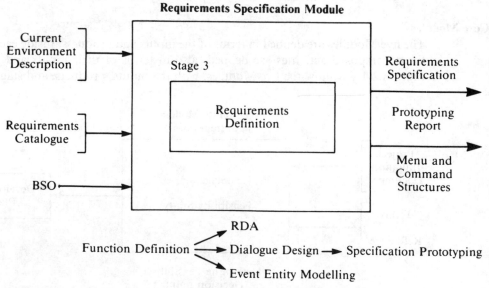

Figure 2.6

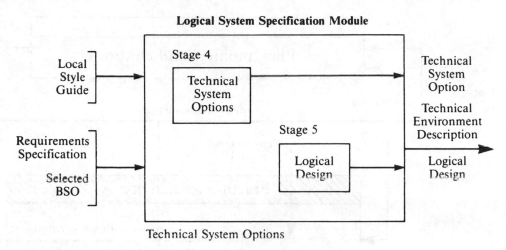

Logical System Specification Module

Stage 4

Technical System Options

Stage 5

Logical Design

Technical System Option

Technical Environment Description

Logical Design

Local Style Guide

Requirements Specification

Selected BSO

Technical System Options

Logical Database Process Design

Figure 2.7

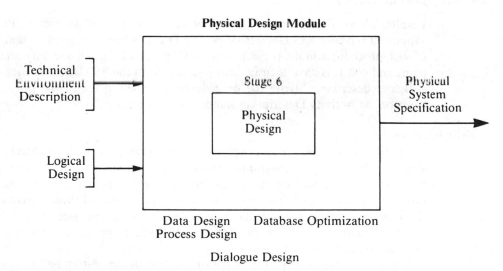

Physical Design Module

Stage 6

Physical Design

Technical Environment Description

Logical Design

Physical System Specification

Data Design   Database Optimization
Process Design

Dialogue Design

Figure 2.8

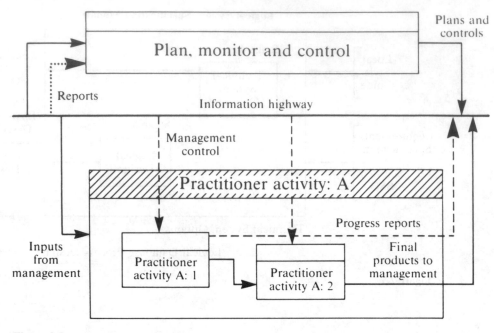

Figure 2.9

2.4   Support activities

I stated above that a second stream of activities, known as Project Procedures, supported the core SSADM activities. SSADM has been designed so that the place of such procedures in the project are well defined, and that all ancillary management plans and controls have their defined interface with the SSADM activities.

Before describing briefly the procedures and their place in the project, I shall describe the Activity Description standards and notation.

Activity interface

'Activity' covers the whole range of core SSADM, from the method, modules, stages, and steps to the contributory tasks that make up each step. Every activity is defined by a particular notation that specifies what happens, what it achieves, and how it is managed. SSADM prescribes the general format of these Activity Descriptions, but each project tailors them, or at least produces project-specific definitions for each. Figure 2.9 portrays the Activity Model that applies to every activity, large or small, undertaken in a project.

The activity itself is depicted by the striped box in the bottom half of the diagram. The project board, or whatever body is governing the project, is represented by the box *plan, monitor and control*. Inside this box, too, are implied all the Project Procedures that are described in Sec. 2.4.

The line marked *information highway*, between activity and project board represents the project manager. The manager is the buffer between the board and the practitioner team performing the activity, and is responsible for passing plans and constraints to the team, and submitting progress and other reports to the board. The

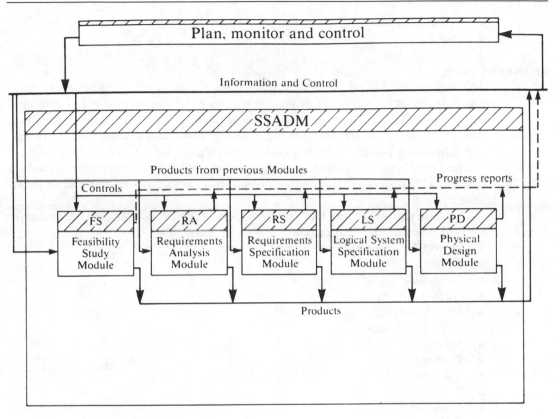

Figure 2.10

end of the activity is the submission of actual products to the information highway at the conclusion.

There are three kinds of flow represented on the diagrams:
- Products
- Progress reports
- Control flows/management authorization

The actual labelling of these flows is the responsibility of Project Management before the start of each activity.

Figure 2.10 uses the above notation to describe SSADM.

The Activity Description that accompanies the diagram describes the following features:
- Objective
- Summary
- Participants
- Preconditions
  —Management authorization
  —Inputs
  —References
- Products

- Techniques
- Tasks

Project Procedures

The philosophy behind the design of SSADM is that it can interface with other, ancillary, activities as part of its normal planning. Any product which is produced outside core SSADM, but must interface with it, is included on the hierarchical Product Breakdown Structure (see Fig. 1.2), and planned into the SSADM process.

All products passing between SSADM and these other activities do so via the information highway. All products that do appear must therefore have been planned for, and will appear as defined inputs or outputs to a SSADM activity at a predetermined point.

There are many activities that fall outside the scope of SSADM, yet are necessary for it to succeed. Principal ones are:

- Project Management
- Planning
- Estimating
- Quality Assurance
- Risk Assessment
- Capacity Planning
- Testing
- Training
- Take-on
- Technical authoring
- Standards

There are many others that a project will need to draw on, particularly if it has peculiar technical requirements such as distribution or real-time facilities. Both of these can be handled by SSADM, but the method will need to be tailored to accommodate the requirements of the environment.

These activities support the SSADM practitioner; the information highway is responsible for passing the method's requirements to these procedures and these products into SSADM, so 'hooks' must be provided for handling these interfaces.

Table 2.1 lists these Project Procedures, whether they provide input to SSADM, receive an output, or both, and, finally, with which part of SSADM they interface (i.e. the 'hook'). This hook may be a generic activity, such as Module or stage, or it may be a specific point, such as Technical System Options.

Chapter 20 describes the activities and requirements of some of these procedures in more detail.

CCTA publications

The emphasis on this second stream of activities interfacing with SSADM means that within the system supplier's organization there must be both expertise in these procedures, and standards to apply to them. Many organizations taking up SSADM may not have standards for some of the activities, although if they build IS systems there must be a certain level of expertise.

As a support to SSADM, CCTA have produced a series of publications for the Project Procedures as part of the method documentation. These describe, but do not

Table 2.1

Procedure	Interface		Level
Project Management Methods	I	O	Modules, stages, via products/plans/structure
Standards	I		Modules, via installation standards
Quality Control	I	O	Modules, stages, via products/descriptions
Capacity Planning	I	O	Feasibility, BSO, TSO, via option descriptions
Risk Assessment	I	O	Feasibility, BSO, TSO, Physical Design, via option descriptions, data descriptions, Function Descriptions
Take-on		O	TSO, Physical Design, via TSO products, and Physical Design products
Technical authoring		O	BSO, TSO, Physical Design, via user Requirements, option description
Testing		O	TSO, via TSO/User Requirements
Training		O	Feasibility/BSO/TSO/Physical Design, via option description, design products

prescribe, particular ways of performing the activities. In some cases, such as Risk Assessment and Management, they recommend a proprietary method; in others, such as Estimating and Metrics, they describe principles and aids instead.

Summary

SSADM is designed with two streams of activity operating: SSADM techniques, in their appointed place, and supporting techniques, Project Procedures, which provide inputs to SSADM, or receive SSADM products of progress reports.

The method is hierarchic, made up of Modules, which are decomposed into stages, which are made up of tasks, which in their turn make use of specific techniques.

There are five Modules:
- Feasibility
- Requirements Analysis
- Requirements Specification
- Logical System Specification
- Physical Design

The output of each of these levels of activity is a set of products, predefined, and to specified quality criteria. The product orientation of the method makes Project Management more reliable, in that there is a measurable deliverable from every activity, rather than a bland statement of progress which, for whatever reason, may be misleading.

The Project Procedures are set in place by Project Management. The procedures are described in a set of publications provided by CCTA. Some of these describe proprietary methods in such areas as Project Management and Risk Assessment and Management, but the successful use of SSADM lies in employing the procedures according to a consistent house standard, rather than in following a specific product.

3. Feasibility Study

3.1 Aims of chapter

The first part of this chapter describes the structure of the Feasibility Study Module. I shall describe the objectives and products of the stage that makes up this Module.

The second part is a breakdown of the procedures that are followed in the stage, detailing the activities carried out in all of the component steps.

3.2 Structure of the Feasibility Study Module

The Feasibility Study Module is carried out in one stage, Stage 0, at the outset of the project. It is triggered by the Project Initiation Document, which itself may be the result of an IS strategy document.

This IS strategy will already have indicated a direction for the project; as the project itself may not be undertaken for some time after the strategy has been published, the technical feasibility and business case must be re-examined, and a range of Business System Options (BSOs) and Technical System Options (TSOs) prepared.

The Feasibility Study is one of the checkpoints where either the project may be aborted, or a contract may be placed for a detailed study to be undertaken. Figure 3.1 shows the Feasibility Study Module's position at the start of the project, triggered by the Project Initiation Document (PID).

Figure 3.2 illustrates the structure of Stage 0, Feasibility. All steps must be performed sequentially and in full.

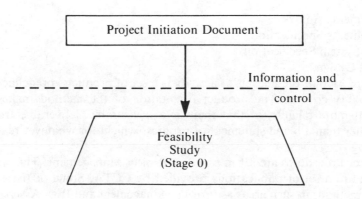

Figure 3.1

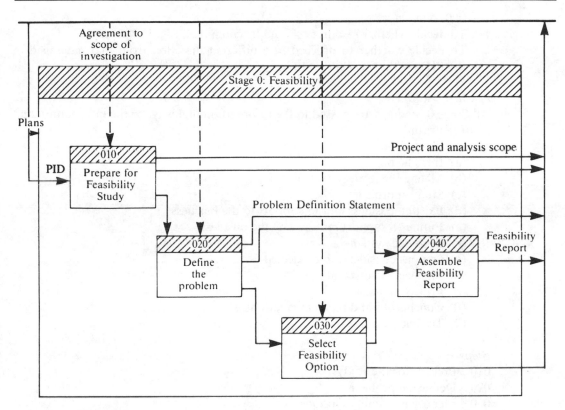

Figure 3.2

Feasibility

In this stage we carry out an assessment of a proposed IS. We are trying to determine whether the system can, in fact, meet the business requirements it is intended to meet, and whether there is a sound business case for developing it. At the end of Feasibility, we should be able to say either that a full SSADM study should be undertaken, or that we should proceed in a different direction.

Ideally, most projects should begin with a Feasibility Study. The exceptions should be low risk projects, which can be aborted, if necessary, at Business System Options.

In many organizations, in fact, the corporate strategy, or business plan, may preclude the need for such a study, and internal politics may move the project straight on to Requirements Analysis. If a third-party provider is bidding for a development contract, the work that goes into the preparation of the bid may well serve the purpose of a Feasibility Study; if the provider should decide on balance not to submit a bid, that is equivalent to a Feasibility Study's recommending the project be aborted. In such a case, it is up to the would-be client company to make the decision.

Feasibility should not be an expensive or time-consuming exercise: one team working for one to two months should be sufficient to produce the possible options and guidance for the way forward.

The study will involve the use of SSADM techniques to examine both the current system and the requirements of the new.

STAGE 0: OBJECTIVES
1. To decide whether resources should be committed to a Full Study.
2. To decide whether to proceed in a different direction from that envisaged in the PID for the Feasibility Study.

STAGE 0: PRODUCTS
All Stage 0 products are passed to the information highway, or the information and control stream.
1. Feasibility Report:
   (a) Introduction
   (b) Management/executive summary
   (c) Study approach
   (d) Existing business and IS support to the business
   (e) Future IS support required by the business
   (f) Proposed system
   (g) Options considered but rejected
   (h) Financial assessment
   (i) Project plans
   (j) Conclusions and recommendations
   (k) Technical annexes

Steps
010 Prepare for the study
020 Define the problem
030 Identify Feasibility Options
040 Assemble Feasibility Report

3.3 Stage procedures

The remainder of this chapter describes the tasks performed in each of the steps that make up this stage.

Step 010 Prepare for the Feasibility Study
In this step we examine the PID and related documentation with the primary objective of agreeing the scope of the study, and planning the rest of the Module's work.

The requirements of the study will be reviewed, with the terms of reference, description of the business environment and culture, the technical environment, and business requirements/problems related to the project.

Before the project moves forward to Step 020 the project board must have agreed the scope and terms of reference, and resolved all questions and problems arising.

Techniques employed in Step 010 are:
- Data Flow Modelling
- Logical Data Modelling
- Requirements Definition

Figure 3.3 illustrates Step 010.

- Project Initiation Document
  ↓

Tasks   *Description*
10   Review the PID and relevant background material. Assess the scope and complexity of the proposed IS.
          Create a Context Diagram, current Level 1 DFD and Overview LDS.
          Identify the base requirements from the PID and enter them in the Requirements Catalogue. Report errors or inconsistencies in the PID.

20   Identify the stakeholders for the business area concerned. Establish how they are to be involved, and brief the User representatives.
          Identify those areas to be investigated, and define the methods to be used.
          Agree the Feasibility Study scope with the project board.

30   Develop an Activity Network, Activity Descriptions, Product Flow Diagrams, Product Breakdown Structure, and Product Descriptions. Agree these with the project board.

    ↓

- Context Diagram
- Current Physical DFD
- Overview LDS
- Requirements Catalogue
- Study plan
- Activity Network
- Activity Descriptions
- Product Flow Diagrams
- Product Breakdown Structure
- Product Descriptions

Figure 3.3

Step 020   Define the problem

In this step we carry out a more detailed investigation of the business and its information needs. We define the Users of the system, the new services wanted from the system, and any problems associated with the current services and operations.

Techniques employed in Step 020 are:

- Data Flow Modelling
- Dialogue Design
- Logical Data Modelling
- Requirements Definition

Figure 3.4 illustrates Step 020.

- Context Diagram
- Current Physical DFM
- Overview LDS
- Requirements Catalogue

↓

Task	Description
10	Identify the activities and information in the area of study that are necessary for the business unit to meet its objectives. Draw a level 1 DFD for the required environment. Amend the Overview LDS to include entities and accesses in the required environment.
20	Investigate the current environment. If necessary, expand DFD to Level 2. Identify, with the User's help, those aspects of the current operations where improvement or change is required. Enter these in the Requirements Catalogue.
30	Define the Users of the new system in the User Catalogue
40	Identify with the Users new features in the required system (i.e. new processes or functions). Record these in the Requirements Catalogue. Identify any non-functional requirements (i.e. service levels, response times, recovery, and security). Record these in the Requirements Catalogue.
50	Prepare a Problem Definition Statement to summarize the requirements, and assess priorities for the business objectives.
60	Agree the Problem Definition Statement with the project board.

↓

- Outline Current Environment Description
- Outline Required Environment Description
- Problem Definition Statement
- User Catalogue

Figure 3.4

Step 030   Select Feasibility Options

In this step we develop a set of options for solutions to the problems identified and agreed. The options are presented to project board, for the selection of one which will define the way forward. The selected option will define one or more projects to meet the requirements of the target business area.

The options combine both business option elements, to define the scope of the project (2), and technical options, to describe the physical environment in which it will operate.

After the board has made its selection, development plans are produced, in outline, for associated projects.

Techniques employed in Step 030 are:
- Business System Options
- Technical System Options
- Data Flow Modelling
- Logical Data Modelling

Figure 3.5 illustrates Step 030.

- Outline Current Environment Description
- Outline Required Environment Description
- Problem Definition Statement
- Requirements Catalogue
- User Catalogue

↓

Tasks	Description
10	Draw up a list of the minimum requirements for the new system, both functional and non-functional. All options must satisfy these basic requirements.
20	Define up to six Business System Options, which will satisfy a range of requirements, from the basic minimum to all listed.
30	Define a list of outline Technical System Options, to represent a full range of technical solutions. Each technical solution should satisfy the constraints and requirements of at least one Business System Option. In consultation with the Users, draw up a list of up to six composite Business and Technical Systems Options. Reduce these to a short list of about three.
50	Describe, in prose, each short-listed option. This description can be supported by DFDs and an LDM, to illustrate the differences between them. They should be supported also by outline Cost/Benefit Analysis and Impact Analysis.
60	Identify the preferred option. Produce an outline development plan for each recommended project.
70	Present the short-listed options to the project board, and other User audiences. Assist in the selection of the one chosen option, with explanation and clarification of the implications of each. Record any decisions made, with reasons.
80	Develop an Action Plan for the selected project(s). This should include a description of the technical approach, and outline development plans.

↓

- Feasibility Options
- Action Plans

Figure 3.5

Step 040   Assemble the Feasibility Report

In this step we publish the Feasibility Report, having checked the accuracy and integrity of the products of the study.

Each SSADM product is subject to a Quality Control check, but that is built into the Project Procedures rather than into an SSADM step. Ensuring integrity and consistency between the products is part of core SSADM. That happens in this step, and the corresponding step of each module.

Each Product Description carries a list of quality criteria, which serve as guides to the checks carried out here.

Figure 3.6 illustrates Step 040

- Action Plan
- Feasibility Options
- Outline Current Environment Descriptions
- Outline Required Environment Descriptions
- Problem Definition Statement
- Requirements Catalogue
- User Catalogue

↓

Task	Description
10	Check the completeness and consistency of the Feasibility Study Module by reviewing the products above. Amend the products as a result of reviews, if necessary or appropriate.
20	Assemble and publish the Feasibility Report documents.

↓

- Feasibility Report

Figure 3.6

Summary

Feasibility is an activity that should ideally be undertaken in all but low risk projects. It is a decision point; the decisions possible include that to terminate the project.

SSADM techniques are employed during the study, but they are not the only techniques used.

The options presented to the project board must address several issues: business, organizational, technical, and financial.

SSADM techniques used in Stage 0 are:

- Data Flow Modelling
- Logical Data Modelling
- Requirements Definition
- Dialogue Design
- Business System Options
- Technical System Options

On completion of this Module, if the project board give their approval, we are ready to move into the first Module of SSADM proper: Requirements Analysis.

Figure 3.7 shows the Product Breakdown Structure for the Feasibility Study Module.

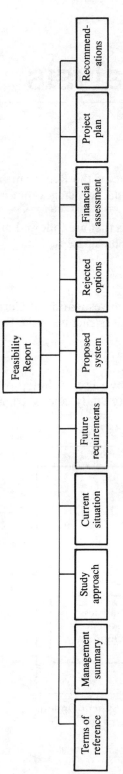

Figure 3.7

31

4. Requirements Analysis

4.1 Aims of chapter

The first part of this chapter describes the structure of the Requirements Analysis Module, as well as the objectives and products of the two stages that make up the Module.

The second part is a breakdown of the procedures that are followed in each stage, detailing the activities carried out in all of the component steps.

4.2 Structure of Requirements Analysis

Requirements Analysis is carried out in two stages: investigation of Current System and Business System Options. Figure 4.1 illustrates the structure of the Requirements Analysis Module.

Investigation of Current Environment

This stage may be the start of the project, triggered by a Project Initiation Document from the project board, or it may follow a Feasibility Study, and be triggered by the Feasibility Report.

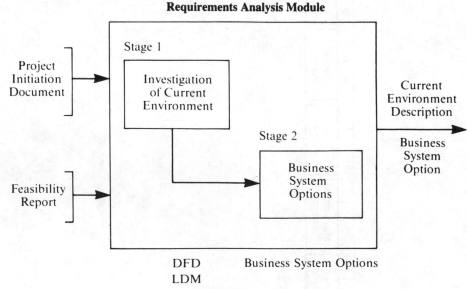

Figure 4.1

However the stage—and hence the analysis and design project—is initiated, it has five objectives to fulfil.

STAGE 1: OBJECTIVES
1. To confirm that the project has the correct brief from management.
2. To prepare the initial task lists and resource estimates for the work.
3. To establish a clear statement of User requirements from the system, both functional and non-functional.
4. To establish roles for the project, especially the position and responsibilities of the Users.
5. To model the procedures and data structures for those areas considered for computerization.

It is apparent that the first four objectives are more appropriate to the Project Management function, and the fifth is the one that corresponds to the traditional analysis stage of a project. The stage, certainly, is a planning stage, with the conduct of the project and the agreement of the project brief having equal importance to the investigation of the business environment.

In most cases, the analysis team will not begin work, or even see the brief, until the first four activities have been carried out, and the project board, or whichever authorizing body is responsible, has given the go-ahead.

STAGE 1: PRODUCTS
1. Passed to information and control:
   (a) Activity Descriptions
   (b) Activity Network
   (c) Product Breakdown Structure
   (d) Product Descriptions
   (e) Product Flow Diagram
2. Passed to Stage 2:
   (a) Requirements Catalogue
   (b) Current Services Description
      (i) Context Diagram
      (ii) EPDs
      (iii) I/O Descriptions
      (iv) Current Environment LDM
      (v) Logical DFDs
      (vi) Logical Data Store/Entity Cross-reference

Steps
110   Establish analysis framework
120   Investigate and define requirements
130   Investigate current processing
140   Investigate current data
150   Derive logical view of current services
160   Assemble investigation results

Business System Options

This stage takes as its input the Current Services Description, the Requirements Catalogue and the Project Initiation Document.

STAGE 2: OBJECTIVES

The objective of the stage is to select a solution to the problems and requirements of the given business environment. The scope of this particular project will finally be settled. Interfaces to other projects and other business areas are agreed during this stage.

The solution is selected from a number put up for consideration by the analysts. The options will consist in part of SSADM products to describe the scope of information processing, and in part of financial, business, and risk assessments to provide a sound business reason for the selection.

STAGE 2: PRODUCTS

Passed to data and control:
(a)   Business System Options
(b)   Selected Business System Option

Steps
210   Define Business System Options
220   Select Business System Options

Stage interfaces

As can be seen from this stage description, each stage and each module interfaces to information and control as well as to the next module. Figure 4.2 illustrates these interfaces for the Requirements Analysis Module.

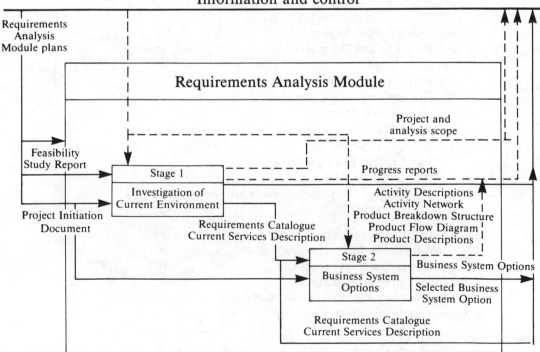

Figure 4.2

4.3   Stage procedures

As described in Chapter 2, each SSADM stage is broken down into steps, and each step is made up of tasks. The remainder of this chapter enumerates the steps and tasks performed in Requirements Analysis. The techniques performed in the tasks are described later in the book (Chapters 8–12). Standard SSADM forms used in these techniques will be described/illustrated at the relevant section of those chapters.

Investigation of Current Environment

Step 110   Establish analysis framework
This step is a concern of Project Management rather than IT analysis, or business analysis.

The Project Initiation Document, containing terms of reference, is examined, together with the results of any earlier study, whether SSADM Feasibility, or other. The inputs to this step will be reviewed to ensure that the project brief is still in line with the business objectives, and that both are consistent with the findings of earlier studies. Should any significant problems be identified here, the project board *must* be consulted before leaving this step.

At the end of this step the necessary project activities and products will be identified. They will form the basis for the Project Plans.

Techniques employed in Step 110 are:
- Data Flow Diagramming
- Logical Data Modelling
- Requirements Definition

Figure 4.3 describes Step 110.

- Feasibility Study Report, or
- Report from previous study
- Project Initiation Document (PID)
  ↓

Task	Description
10	Review the PID and the outputs from previous studies. Review Context Diagram, Current Physical DFD and Overview LDS, from earlier study. Make any changes needed to make them reflect the current environment. Identify specific system requirements from previous studies, and enter them in the Requirements Catalogue. Report any errors or inconsistencies in the input documents and wait until they are resolved.
20	Identify the target Users for the system, and establish their role in the analysis. Brief the User representatives. Identify the areas for investigation, and define the means of investigation (interview, questionnaire, document sampling, etc.) Agree the scope of the project and analysis with the project board.
30	Develop an Activity Network, Activity Descriptions, Product Flow Diagram, Product Breakdown Structure, and Product Descriptions for those SSADM parts of the project. Agree these items with the project board.

↓

- Activity Descriptions
- Activity Network
- Context Diagram
- Current Physical DFD
- Overview Logical Data Structure
- Product Flow Diagram
- Product Descriptions
- Requirements Catalogue

Figure 4.3

Step 120   Investigate and define requirements

In this step we identify the problems in the current environment that are to be resolved, and also identify any additional services that the new system must provide.

The Requirements Catalogue (initiated in the previous step) is expanded in the light of more detailed investigation.

Requirements, under the twin headings of 'new requirements' and 'current problems', are identified in broad terms only, and certainly no solutions are identified yet.

The technique employed in Step 120 is:

- Requirements Definition

Figure 4.4 illustrates Step 120.

- Requirements Catalogue
- User Catalogue
  ↓

Task	Description
10	Investigate the operation of the current system. Record details (for later use) of environment details, e.g., volumes, frequencies, hardware usage.
20	Define the intended Users of the new system in the User Catalogue.
30	Identify from discussion with the Users those aspects of the current system that present an opportunity for improvement. Enter these in the Requirements Catalogue.
40	Identify from discussion with the User those functions and data not yet provided by the current system. Enter these as requirements in the Requirements Catalogue.

↓

- Requirements Catalogue

Figure 4.4

Step 130   Investigate current processing

Step 130 describes the information flows in the environment in the form of DFDs. This activity is carried out in parallel with Step 120 (Investigate and define requirements) and Step 140 (Investigate current data).

The Level 1 DFD from Step 110 is expanded and taken down to Levels 2 and 3.

The technique employed in Step 130 is:

- Data Flow Modelling

Figure 4.5 describes Step 130.

- Context Diagram
- Current Physical Data Flow Diagram
- Requirements Catalogue
- User Catalogue

↓

Task	*Description*
10	If the Level 1 DFD is complex, or unclear, it might help to unpick what is really happening by drawing a Document Flow Diagram for each flow on the Level 1 DFD. If you do do this, then you will need to perform Task 20. Otherwise, begin this step at Task 30.
20	Combine the Document Flow Diagrams from Task 10 into one network. Use this network to improve the Level 1 DFD. If there are discrepancies between the two diagrams (which is probable), resolve them by discussion with the User.
30	Draw a Level 2 DFD for each Level 1 process. Decompose the processes to the lowest level appropriate.
40	Create Elementary Process Descriptions for each lowest level process.
50	Create an Input/Output Description for each lowest level data flow across the system boundary.
60	Identify with the Users any shortcomings in the current processing and record them in the Requirements Catalogue. These shortcomings may be bottlenecks, excessive duplication of data, superfluous procedures; in short, any feature that tends to suboptimum performance.

↓

- Context Diagram
- Current Physical DFDs
- Input/Output Descriptions

Figure 4.5

Step 140   Investigate current data
This step describes the logical structure of the data used in the current system. By 'logical' I mean that the structure of the data is divorced from the physical way it is held, in, for example, card indexes, or price lists, or filing cabinets.

The step is carried out in parallel with Step 120 (Investigate and define requirements). At this point, the data model reflects only the requirements for the current processing.

The model will be validated against the processing documented in the Elementary Process Descriptions.

The technique employed in Step 140 is:
- Logical Data Modelling

Figure 4.6 illustrates Step 140.

- Overview Logical Data Structure
- Requirements Catalogue
- User Catalogue
- Elementary Process Descriptions
                    ↓

Task	Description
10	Create an outline LDS of the current system data. It may be useful to begin with an Entity/Entity Matrix unless fewer than a dozen entities are identified. (As most systems will have far more than a dozen, a matrix is almost a routine requirement. A CASE tool that supports Logical Data Modelling should have at least the equivalent of a matrix, for identifying relationships, and maybe actually chart the structure from the matrix.)
20	Define the important attributes associated with each entity.
30	Informally validate the model against the Elementary Process Descriptions. Each operation in the EPD should be able to access each appropriate data item by navigating the structure. If performed successfully, the next task will not be needed. If there are problems in retrieving any data, then perform Task 40.
40	Identify, with the User's assistance, any deficiencies in the availability of data, and record it in the Requirements Catalogue.

                    ↓
- Current LDS
- Updated Requirements Catalogue

Figure 4.6

Step 150   Derive logical view of current services
This step refines the products of Step 130, by taking the DFDs of the current system and revising them so that they reflect the business logic of the system, rather than the physical implementation.

By identifying only the logical aspects of the processing, some of the problems previously identified will be resolved, such as redundant processing and anomalous procedures and data storage.

Any solutions that are achieved in this way must be recorded against the appropriate entry in the Requirements Catalogue. Other problems and requirements will remain, however, and must not be considered at this point.

The procedures for achieving this are described in greater detail in Chapter 9.

The techniques employed in Step 150 are:
- Data Flow Diagramming
- Requirements Definition

Figure 4.7 illustrates Step 150.

- Context Diagram
- Current Environment LDM
- Current Physical DFD
- I/O Descriptions
- Requirements Catalogue
- User Catalogue

↓

Task	Description
10	Remove traces of physical considerations from Levels 2 and 3 DFDs. Thus, location/role names will vanish from the process boxes, and local names of documents from data flows.
20	Rationalize the data stores, and relate each data store to one or more entities on the LDM. Each entity will then be found in no more than one data store.
30	Rationalize the processes on the lowest level DFDs, and group them together to form a Level 1 Logical DFD. Amend the supporting documentation (EPDs and I/O Descriptions) to reflect the new diagrams.
40	Cross-validate EPDs against the LDM.
50	Update the Requirements Catalogue to reflect those solutions to do with the physical constraints identified in this step.

↓

- Current Environment LDM
- Logical Data Store/Entity Cross-reference
- I/O Descriptions
- Logical Data Flow Diagrams
- Requirements Catalogue
- User Catalogue

Figure 4.7

Step 160   Assemble investigation results

This step completes the investigation of the Current Environment. Its objective is to ensure the consistency and integrity of the products of Stage 1. Quality reviews are held of all the products developed during the stage.

Figure 4.8 illustrates Step 160.

- Current Environment LDM
- Logical DFD
- Logical Data Store/Entity Cross-reference
- EPDs (Current logical)
- I/O Descriptions (Current logical)
- Requirements Catalogue
- User Catalogue

↓

Task	Description
10	Carry out quality reviews of the input products above. Amend the products as necessary in the light of the reviews.

↓

- Current Services Description
- Requirements Catalogue
- User Catalogue

Figure 4.8

Business System Options

Step 210   Define Business System Options
This step addresses the identified requirements by producing a number of possible solutions to them. Each of these solutions will differ to a greater or lesser degree, depending on impact, functionality and cost/benefit.

The technique employed in Step 210 is:
- Business System Options

Figure 4.9 illustrates Step 210.

- Current Services Description
- Project Initiation Document
- Requirements Catalogue
- User Catalogue

↓

Task	Description
10	Draw up a list of the minimum requirements, both functional and non-functional. The minimum will be those and only those marked as 'mandatory' on the Requirements Catalogue.
20	Define up to six possible business solutions to these minimum requirements.
30	Discuss the options with the Users, and cut them down to a short list of three.
40	Describe each short-listed BSO. The description will be text supported by LDM and DFD, illustrating the differences between them. The descriptions must also include an analysis of the organizational impact, and a Cost/Benefit Analysis.

↓

- Business System Options

Figure 4.9

Step 220   Select Business System Option

In this step we present the short list to the project board, and help them to select the one that will be developed. Should the selected option be a mix-and-match affair from those on offer, as often happens, the definition of the option must be amended to cover the new functionality and features.

The technique employed in Step 220 is:

● Business System Options

Figure 4.10 illustrates Step 220.

● Business System Options
    ↓

Task	Description
10	Formally present the BSOs to the project board. They will make a decision, with your assistance, on the most appropriate solution for the specified requirements. Record any decision reached.
20	Complete a description of the selected BSO. This will form the basis for the specification developed in Stage 3.
	If the selected option is exactly as the one represented, then most of the work is done; if the option chosen is a hybrid of two or more selected, a new description should be developed. This will include DFDs, LDS, Cost/Benefit Analysis, Impact Analysis and narrative. Also to be included are reasons for the selection of that option and the rejection of the others.

    ↓
● Selected Business System Option

Figure 4.10

Summary

In Requirements Analysis we have carried out an investigation into the workings of the Current Environment. Using the two views of the data we have gained an understanding of the functional workings of the system, the data model that currently supports it, and a list of the problems and requirements that need to be addressed.

Using these as our starting point, we have now produced a number of solutions to those problems and requirements, and presented them to the User. On the basis of cost/benefit, organizational impact, functionality, or whatever criterion is most important, the User has made a selection from the options. This selection may very likely be a hybrid of two or more of the options presented. Whatever has been chosen by the User must now be developed in sufficient detail to form the basis for design of the new system.

Techniques employed in this Module are:

● Data Flow Modelling
● Requirements Analysis
● Logical Data Modelling
● Business System Options

The products from these activities may now be carried forward into the next Module—Requirements Specification.

Figure 4.11 gives the Product Breakdown Structure for the Requirements Analysis Module.

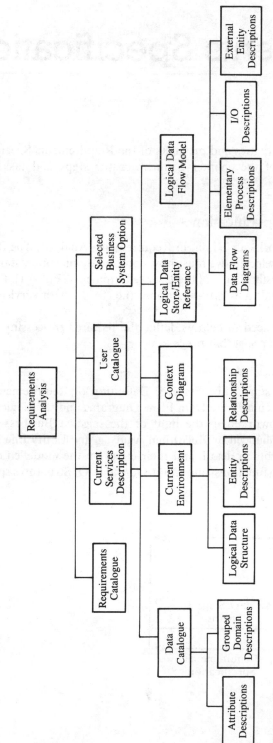

Figure 4.11

43

5. Requirements Specification

5.1 Aims of chapter

This chapter describes the activities and products of the Requirements Specification Module. The second part of the chapter enumerates the steps and tasks to be undertaken.

5.2 Structure of Requirements Specification

This Module develops the work carried out in Requirements Analysis. The Business System Option having been selected, we must now consolidate our understanding—and, indeed, the User's understanding—of the User requirements (Fig. 5.1). Requirements will be both functional, and non-functional (i.e. to do with service levels, security, recovery, etc.).

New techniques are introduced to help us define the required processing, and to define the data structure to support that processing.

Definition of Requirements

This module comprises one stage only, Stage 3: Definition of Requirements. The products from Stage, 1, particularly the Data Flow Diagrams and the Logical Data Model, are reviewed and reworked in the light of the selected Business System Option. The data model is enhanced by Relational Analysis and Entity Life History Analysis, which specifies events in detail, and their effect on the modelled entities. The DFDs are translated into functions, and Input/Output (I/O) Structures specified.

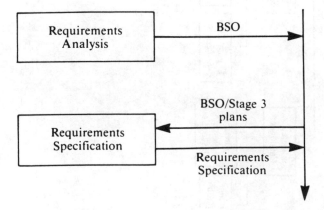

Figure 5.1

STAGE 3: OBJECTIVES
1. To provide for User management's approval a Requirements Specification Document for the development of a Logical System Specification.
2. To specify measurable acceptance criteria for subsequent design products.

STAGE 3: PRODUCTS
All products from this stage are passed to information and control.
(a) Function Definitions
(b) Requirements Catalogue
(c) User Role/Function Matrix
(d) Input/Output Structure
(e) Prototyping Report
(f) Required Systems Logical Data Model
(g) Entity Life Histories
(h) Effect Correspondence Diagrams
(i) Enquiry Access Paths
All of these products are combined in the document from Stage 3, Requirements Specification.

Steps
310   Define Required System processing
320   Develop Required Data Model
330   Derive system functions
340   Enhance Required Data Model
350   Develop Specification Prototypes
360   Develop Process Specification
370   Confirm system objectives
380   Assemble Requirements Specification

5.3   Stage procedures

The eight steps listed above are not strictly sequential. Steps 310 and 320 are carried out in parallel. Step 310 feeds into 330, and Step 320 into 340. Step 350 may not begin until 310 and 330 are complete, while 330 and 340 are required to trigger 360.

Figure 5.2 shows the structure of the stage.

Definition of Requirements

Step 310   Define Required System processing
Step 310 is carried out in parallel with Step 320. The Logical DFDs are fully amended to reflect the Selected BSO. Just as in the current description, where the overview is expanded and decomposed to the lowest level, we carry out the same operation of the Level 1 DFD for the new system.

The contents of all data flows across the system boundary must be defined in full at this point, so that they can be fed into subsequent Function Definitions. A dictionary must also be built up of all data stores, and their data items. This will form our Data Catalogue.

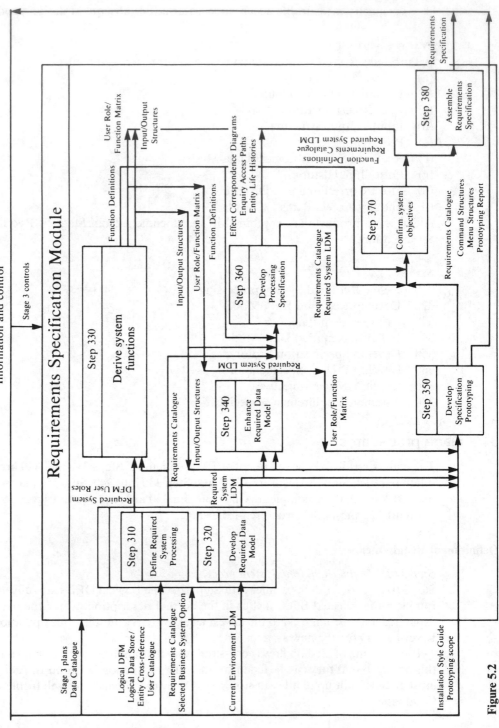

Information and control

Requirements Specification Module

Stage 3 controls

Step 330
Derive system functions

Step 360
Develop Processing Specification

Step 370
Confirm system objectives

Step 380
Assemble Requirements Specification

Step 340
Enhance Required Data Model

Step 350
Develop Specification Prototyping

Step 310
Define Required System Processing

Step 320
Develop Required Data Model

Stage 3 plans
Data Catalogue

Logical DFM
Logical Data Store /
Entity Cross-reference
User Catalogue

Requirements Catalogue
Selected Business System Option

Current Environment LDM

Installation Style Guide
Prototyping scope

User Role/
Function Matrix

Input/Output
Structures

Function Definitions

Input/Output Structures

User Role/Function Matrix

Function Definitions

Effect Correspondence Diagrams
Enquiry Access Paths
Entity Life Histories

Function Definitions
Requirements Catalogue
Required System LDM

Requirements Catalogue

Input/Output Structures

Required System LDM

User Role/Function Matrix

Required System
DFM User Roles

Required
System
LDM

Function Definitions
Requirements Catalogue
Required System LDM

Requirements
Specification

Requirements Catalogue
Command Structures
Menu Structures
Prototyping Report

Figure 5.2

46

During this activity, we should also identify which User Roles are responsible for initiating which processes, and performing which tasks. This will help us further down the road in both Function Definition and Dialogue Design.

Techniques employed in Step 310 are:
- Data Flow Modelling
- Requirements Definition

Figure 5.3 illustrates Step 310.

- Logical Data Flow Diagrams
- Requirements Catalogue
- User Catalogue
- Selected BSO
- Logical Data Store/Entity Cross-reference
- Elementary Process Descriptions

↓

Task	Description
10	Redraw the Level I Current DFD to come into line with the Selected BSO. Add any new processes identified in the BSO, and remove processes no longer inside the system boundary.
20	Amend the lower level DFDs to support any new processing requirements. Update the Requirements Catalogue to include reference to the new or amended processes.
30	For each new lowest level DFD create an Elementary Process Description. Where appropriate, amend existing EPDs. For each lowest level data flow crossing the system boundary, create new I/O Descriptions; amend existing descriptions where appropriate.
40	Validate the data stores on the required model against the Logical Data Model: make sure that each store consists of one or more entities. Ensure that the attributes listed against an entity are consistent with the data flows to and from the data store.
50	Define the User Roles in the required system, and ensure that these roles can be mapped on to the external entities on the required DFD.

↓
- User Roles
- Amended Requirements Catalogue
- I/O Description (required system)
- EPDs (required description)
- Logical Data Store/Entity Cross-reference

Figure 5.3

Step 320   Develop Required Data Model

This step is carried out in parallel with Step 310. The Logical Data Model of the Current Environment is amended to support the requirements defined in the Requirements Catalogue. It is important that for this version, all attributes are identified and recorded in detail.

Techniques employed in Step 520 are:

● Logical Data Modelling
● Requirements Analysis

Figure 5.4 illustrates Step 320.

● Current LDM
● Requirements Catalogue
● Selected BSO

↓

Task	Description
10	Review the Selected BSO, and amend the Current LDM to take account of the additional requirements. There will be some decisions to make that are more design-oriented, to do with, for example, use of entity subtypes and supertypes.
	The larger the LDS, the more possible ways of representing data and dependencies there may be. However the model is drawn, the analyst must remember always that the business rules must be represented. The structure forces some of these rules, such as optionality or exclusive relationships; others, to do with, for example, classifying the data, will probably need to be resolved with the help of the Users.
	Complete the Entity, Relationship and Attribute Descriptions.
	Amend the Requirements Catalogue to reference the changes made for each requirement, showing where this provides solutions to the requirements.
20	Validate the EPDs with the extended model.

↓

● Required System Data Model
● Amended Requirements Catalogue

Figure 5.4

Step 330   Derive system functions

In this step we identify functions for both update and enquiry. For this we use the Required DFDs and the Requirements Catalogue.

Dialogues are specified—but not designed—and those critical to the system's success are identified.

Techniques employed in Step 330 are:

● Function Definition
● Dialogue Definition
● Requirements Definition

Figure 5.5 illustrates Step 330.

- Requirements Catalogue
- User Roles
- Required DFDs
- Required I/O Descriptions
- Required EPDs
- Logical Data Store/Entity Cross-reference

↓

Task	Description
10	With the Users, identify the update functions from the DFDs. Every lowest level DFD must have at least one function allocated to it. Identify the events that trigger each function. There may be just a one-to-one correspondence for events and functions in many cases, but some functions may be triggered by more than one event.
20	With the Users, identify all enquiry functions.
30	For each function, specify the I/O interface, using the I/O Descriptions from the DFD for the update functions. Consult with the Users for specification of the enquiry interfaces.
40	Cross-reference the User Roles and Functions, to identify the required system dialogues. Identify which dialogues are critical to the system.

↓

- I/O Structures
- Function Definitions
- Requirements Catalogue
- User Role/Function Matrix

Figure 5.5

Step 340   Enhance Required Data Model
In this step, we validate and enhance the LDM by the use of Relational Data Analysis (RDA). The source of this analysis is the set of I/O. Descriptions from Step 330. The normalized relations from this exercise are used to build submodels, which we then compare to the LDM.

   Techniques employed in Step 340 are:
- Relational Data Analysis
- Logical Data Modelling

Figure 5.6 illustrates Step 340.

- I/O Structures
- Required LDM

↓

Task	Description
10	Select those functions whose I/O Structures will be subject to RDA. Not all I/Os will be used, but those that are selected should, between them, cover the data content of the model.
20	Carry out Relational Data Analysis to Third, or Boyce–Codd, Normal Form, on the specified I/O Structures, and produce a set of normalized relations for each function.
30	Convert each set of normalized relations into a data submodel.
40	Compare the submodel with the corresponding part of the LDS. Resolve any discrepancies with the Users, and by reference to the Requirements Catalogue.

↓

- Required System Logical Data Model

Figure 5.6

Step 350   Develop Specification Prototypes
In this step we develop prototype models of selected parts of the specification in order to validate that specification, and to ensure that the requirements are fully understood. This is not an incremental prototype, and the Users should be clear before the exercise begins that they are seeing only a model, not the finished product. If this explanation is carried out badly by the analysis team, User expectations can be built up wrongly, and therefore let down painfully.

   Techniques employed in Step 350 are:
- Specification Prototyping
- Requirements Analysis
- Dialogue Definition
- Requirements Analysis

Figure 5.7 illustrates Step 350.

- I/O Structures
- Requirements Catalogue
- User Role/Function Matrix
- Installation Style Guide
- Prototyping scope

↓

Task	Description
10	Using the prototyping scope (from Project Management) select the dialogues and reports to be prototyped.
20	Create prototypes of the Dialogue Menus and Command Structures for the specified User Roles. If possible, the prototyping tool should be the same as the implementation platform. This is not always possible, depending on local circumstances and policies, but try at least to simulate the platform on which the final system will run. Demonstrate the menu prototypes to the holder of the appropriate User Role. Modify and reshow until they are accepted.
30	Identify the screen and report components that are to be prototyped. Combine these with the approved Dialogue Menus to create the complete pathway that the Users will follow in their dialogues.
40	Implement the pathways.
50	Prepare for prototyping sessions with the Users.
60	Carry out the prototyping sessions with the nominated Users.
70	Review the sessions, and report the results. Steps 40–70 may be repeated for each of the pathways being prototyped.
80	Review the results of the prototyping, and note any errors that have been identified in the Requirements Specification. Amend the Requirements Catalogue with details of the User interface requirements established during the prototyping sessions. Complete the report for management on the sessions.

↓

- Prototyping Report
- Menu and Command Structures
- Amended Requirements Catalogue

Figure 5.7

Step 360   Develop Processing Specification

In Step 360 we define the processing requirements (for update only), expanding on the information gained from DFDs. We use a new tool, Entity/Event Analysis, that gives a third view of the system data in that it shows the effect of time on our entities. We identify the events that impact on our system, and model their effects for every entity. The business rules prescribe the sequence in which these events will affect each entity.

Entity/Event Modelling combines with Function Definition, Logical Data Modelling and Data Flow Modelling in an iteration of further discovery about the system. If a new event is discovered during this step, then perhaps a new process on the DFD and so a new Function Definition will also be required. This process of further understanding can go round in a number of passes, rather like a spiral around the functionality of the system, until analysts and Users are satisfied that everything is fully understood and modelled.

The techniques employed in Step 360 are:

- Entity/Event Analysis
- Function Definition
- Requirements Definition

Figure 5.8 illustrates Step 360.

- Function Definitions
- I/O Structures
- Required LDM
- Requirements Catalogue

↓

Task	Descriptions
10	For each entity on the LDM identify all events that update it, whether creation, modification or deletion. Work bottom-up from the LDS. The Function Definitions will have identified many of the events. For every event identified, create an Event Specification. For every new event identified, define the functions and amend the Functions Definitions for redefined events.
20	Create a 'first-cut' Entity Life History for each entity on the LDS. For this first cut draw only a 'normal life'. After all have been drawn, include any 'parallel life' events.
30	Working top-down on the LDS, refine the ELHs to include abnormal death events, and any interaction between entities.
40	Add operations to the ELHs.
50	Amend the Requirements Catalogue with any new requirements identified in the course of ELH analysis.
60	Draw an Effect Correspondence Diagram for each event.
70	Create an Enquiry Access Path for each Enquiry Function.

↓

- Entity Life Histories
- Effect Correspondence Diagram
- Event Specification
- Function Definition
- Requirements Catalogue
- Enquiry Access Paths

Figure 5.8

Step 370   Confirm system objectives

In this step, we re-examine the Requirements Catalogue to ensure that the identified requirements are met in the specification. Not only should the requirements be addressed, but measures must also be defined for assessing how well the requirements are met.

In addition to these tasks, we must examine the non-functional requirements to ensure that these are defined. Such requirements will be in the nature of operational requirements or constraints, or such considerations as audit requirements. This is an important point in the project, in that it marks the end of the analysis phase, and the start of purely design activities. It is important, therefore, that the specification is both full and accurate.

In many cases, a fresh team may be put onto the design activities or, in cases of third-party suppliers, a fresh contract may be put out to tender for the design work. In these circumstances, with the team responsible for specifying the new system now out of the picture, the importance of this step is obvious.

Techniques employed in Step 370 are:
- Functions Definition
- Requirements Definition

Figure 5.9 illustrates Step 370.

- Function Definitions
- Required System LDM
- Requirements Catalogue

$\downarrow$

Task	Descriptions
10	Examine the Requirements Catalogue to ensure that each functional requirement is fully defined. The definition should include such features as service levels and test criteria. After each definition is checked, ensure that each requirement is satisfied in the new specification.
20	Identify any non-functional requirements that are not yet defined. Ensure that all non-functional requirements are defined and addressed.

$\downarrow$

- Function Definitions ⎫
- Required LDM   ⎬ Amended
- Requirements Catalogue ⎭

Figure 5.9

Step 380   Assemble Requirements Specification

In this Step, we assemble the documents for the Requirements Specification and subject them to a review. SSADM products are each given quality checks, but there is little attention given to the integrity of products passed between steps. Step 380 is one place in the method where such integrity checks are made.

Once the reviews are completed, the formal document is published with all relevant end-of-module reports for management.

Figure 5.10 illustrates Step 380.

- Effect Correspondence Diagrams
- Entity Life Histories
- Event Specification
- Enquiry Access Paths
- Function Definitions
- Input/Output Structure
- Required LDM
- Requirements Catalogue
- User Role/Function Matrix

↓

Task	Description
10	Carry out reviews of the products that form the input to this step, ensuring completeness and consistency. Make any amendments to the documentation or products that the review process indicates.
20	Assemble and publish the Requirements Specification document.

↓

- Requirements Specification

Figure 5.10

Summary

In this module we have developed the Requirements Specification, based on the Business System Option selected in the previous stage.

We have deepened our knowledge of the environment by the use of Event/Entity Analysis, and Relational Data Analysis; we have broadened our understanding of the requirements from the new system by the use of prototyping with the User, and by defining fully the functions to be supported by the system.

Techniques employed in this Module are:

- Relational Data Analysis
- Entity Life History Analysis
- Specification Prototyping
- Function Definition
- Requirements Definition

Having completed this module, we are now ready to move forward to develop this specification in the first part of the design activities: Logical System Specification.

Figure 5.11 shows the Product Breakdown Structure for the Requirements Specification Module.

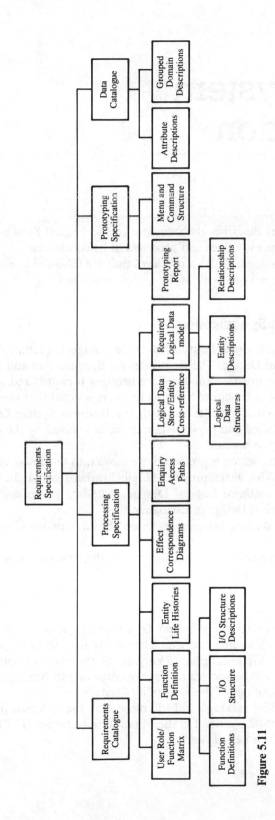

Figure 5.11

6. Logical System Specification

6.1 Aims of chapter

The first part of this chapter describes the structure of the Logical System Specification Module, as well as the objectives and products of its two stages.

The second part is a breakdown of the procedures that are followed in each stage, and details the activities carried out in all of the component steps.

6.2 Structure of Logical System Specification

Logical System Specification is carried out in two stages: Stage 4, Technical System Options and Stage 5, Logical Design. The Requirements Specification and Business System Option from the two previous modules are brought forward, and solutions are found on two fronts: first, a Technical System Option is selected that specifies the technical and development environments. As with the Business System Option, a number of options are produced and presented, and one is selected by the Users as being most appropriate to their needs.

After the Technical System Option is produced, we move into the design activities. At this point we are still looking at the project in logical rather than physical terms. Activities carried out here embrace Logical Design of update processes, Logical Design of enquiries and Logical Design of dialogues.

Figure 6.1 illustrates the interface between the three Modules discussed so far and the data and control highway.

Figure 6.2 presents the Activity Diagram for the Logical System Specification Module.

Technical System Options

In this stage we devise up to six Technical System Options to implement our business solution, and present them to User management. They will make a choice—perhaps a hybrid of several features of different options. Very likely, the range of options will be much fewer than six, and will be constrained by the corporate IS strategy, existing procurement policy, or the findings of the Feasibility Study.

In bid situations, the supplier making a bid will often have the technical platform, hardware, software, and DBMS, specified in their invitation to tender (ITT). In this case, the TSO activities will not be followed.

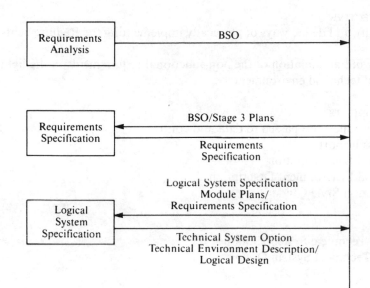

Figure 6.1

Information and control

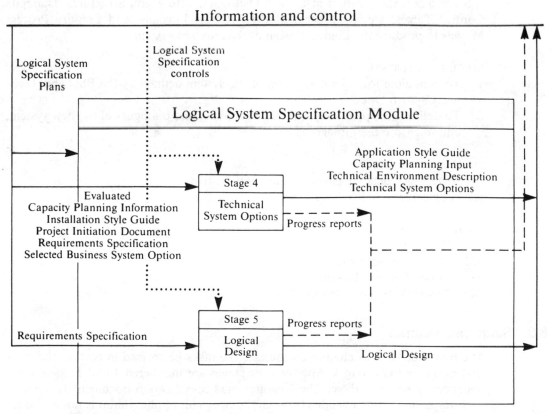

Figure 6.2

STAGE 4: OBJECTIVES
1. To identify and define ways of physically implementing the Requirements Specification.
2. To carry out a validation of the non-functional requirements in the light of the proposed technical environment.

STAGE 4: PRODUCTS
All Stage 4 products are passed to data and control:
(a) Progress Reports
(b) Technical System Option
(c) Technical Environment Description
(d) Application Style Guide

Steps
410   Define Technical System Options
420   Select Technical System Options

Logical Design

This stage is carried out in parallel with Stage 4: there is no dependency between them, and so no requirement to wait for Technical System Options to be completed.
   Stage 5 sees the creation of Logical Dialogues, with Menu Structures, Dialogue Control Tables, Update Process Models, Report Formats, and Enquiry Process Models to produce the Logical Design of the required system.

STAGE 5: OBJECTIVES
1. To create the logical specification of the system defined by the BSO and Technical System Option.
2. To define the Update and Enquiry Processing and dialogues of the new system, and to ensure their integrity.

STAGE 5: PRODUCTS
All Stage 5 products are passed to data and control:
(a) Logical System Specification

Steps
510   Design User Dialogues
520   Define Update Processing
530   Define Enquiry Processing
540   Assemble Logical Design

6.3   Stage procedures

The remainder of this chapter enumerates the tasks performed in each of the steps that make up these stages. Although the stages are numbered 4 and 5, there is no dependency between them. The Requirements Specification document is input to each stage, and is not amended between the two. This implies that it is not necessary to wait for Stage 4 to be completed before beginning work on Logical Design. Since much of the work in Technical System Options may involve waiting for information from suppliers or technical support staff, it is useful for Project Management margins to be able to carry out parallel activities like this.

Technical System Options

Step 410   Define Technical System Options
In this step we identify several possible ways of implementing the Requirements Specification. We also confirm and validate the non-functional (service) requirements in the light of our specified technical environments.

We will probably be constrained in our choice of environments as a result of corporate IS strategy and/or Feasibility. Whatever constraints (or requirements) apply should be a part of the Requirements Catalogue.

The products of this step will be up to six documented options for presentation to management. Included in the documentation for each will be criteria such as Cost/ Benefit and Impact Analysis to help management take an informed decision on the course to follow.

The technique employed in Step 410 is:
* Technical System Options

Fig. 6.3 illustrates Step 410.

* Requirements Specification
* Selected Business System Option
* Project Initiation Document
* Evaluated Capacity Planning information

↓

Task	Description
10	Examine all the input documents and identify the constraints that the options must meet.
20	Define a number—not greater than six—of technical solutions to the requirements.
30	Discuss these options with the Users, and discard any that they regard as non-starters. This should leave you with a short list of two or three.
40	Expand each surviving option into a full description, including for each: • Technical Environment Description • System Description
50	For each option, assess the Capacity Planning information. The service-level requirements must be met for each option, or else the predicted variances must be described in the Technical Environment Description.
60	For each option, add: • Impact Analysis • Cost/Benefit Analysis • Outline Development Plan

↓

* Technical System Options
* Technical Environment Descriptions

Figure 6.3

Step 420   Select Technical System Options
In this step, we present our short-listed options to the project board, and help them to make a selection. The decision is then documented, and the context for Physical Design so defined.

There should be no surprises for the board at the presentation, as we will have discussed all the possible options with them before agreeing the short-list. Even so, they may not choose one option as it stands, but 'mix and match' from the different ones on offer. If their choice is such a hybrid, then it too must be analysed, described, and documented as the others were.

The technique employed in Step 420 is:
- Technical System Options

Figure 6.4 illustrates Step 420.

- Technical System Options
- Installation Style Guide

↓

Task	Description
10	Make a presentation to the board of each option. Help in the decision-making, and note the reasons for the decision.
20	Amend the selected option if appropriate. Develop the full Technical Environment Description.
30	Ensure that the service-level requirements will still be met. The tool for this is Capacity Planning. If the requirements cannot be met in full, this must be agreed with the User, and the Requirements Catalogue annotated accordingly.
40	Develop an Application Style Guide, based on the installation's own standards.

↓

- Selected Technical Option
- Technical Environment Description
- Application Style Guide

Figure 6.4

Step 510   Design User Dialogues
In this step we design the Logical Dialogue for the system. Note that it is not the physical dialogue or physical screens that we are designing, but the logic of the exchange of data. A part of this work may well have been completed in Specification Prototyping, in Step 350, but the bulk of it is still to be done.

The techniques employed in Step 510 are:
- Dialogue Design
- Requirements Definition

Figure 6.5 illustrates Step 510.

- I/O Structures
- Function Definitions
- Requirements Catalogue
- User Role/Function Matrix
- Application Style Guide

↓

Task	Description
10	Specify the syntax checks needed for input data.
20	Identify the logical grouping of elements in the dialogue; the basis for this will be the I/O Structures from Step 330, Function Definition.
30	Identify the navigation paths in each dialogue. Complete the Dialogue Control Table.
40	Define a menu set for each User Role. Define valid control paths on completion of each dialogue. Again, some of this will have been completed during the prototyping activity.
50	Agree and define the 'help' requirements for each level of dialogue.

↓

- Dialogue Structures
- Menu and Command Structures
- Dialogue Control Tables
- Requirements Catalogue

Figure 6.5

Step 520   Define Update Processing

In this step we specify the logic of each database update required for an event. The required updates were defined in Stage 3, Definition of Requirements. Now we consolidate the entity updates into a single process structure for each event.

The ELHs from Stage 3 are completed with the addition of state indicators, and then the Effect Correspondence Diagrams are developed into a processing structure for that event. Semantic validation, using the state indicators, is defined for each event.

The techniques employed in Step 520 are:
- Logical Database Process Design
- Entity/Event Analysis

Figure 6.6 illustrates Step 520.

- Function Definitions
- Event Specification
- Entity Life Histories
- Effect Correspondence Diagrams
- Input/Output Structures
- Required System LDM
- Application Style Guide
  ↓

Task	Description
10	Allocate state indicator values to the ELHs.
20	Convert the ECDs into a processing structure, one per event.
30	With reference to the ELHs, list the operations for each entity affected by the event. Remove reference to gains and losses, and add navigation operations, with further operations for 'fail if . . .' semantic checks.
40	Allocate the operations to the processing structure. Specify conditions governing selections and iterations.
50	Design formats for inputs and reports for the off-line parts of update functions.

↓

- ELHs
- Update Processing Models
- Entity Descriptions
- Report Formats

Figure 6.6

Step 530   Define Enquiry Processing

In this step we complete the Logical Specification of all Enquiry Processing, as identified at Stage 3. We take the I/O Structures and Enquiry Access Paths already identified, and for each develop a single process structure. The structure is derived by merging an input structure (the Enquiry Access Path), and an output structure (the I/O Descriptions, with input elements removed). Operations and validations are added to this new structure. Navigations, retrievals, and semantic checks make up the operations list for these structures.

The technique employed in Step 530 is:

- Logical Database Process Design

Figure 6.7 illustrates Step 530.

- Input/Output Structures
- Enquiry Access Paths
- Entity Life Histories
- Entity Descriptions
- Function Definitions
- Required System LDM
- Application Style Guide

↓

Task	Description
(Steps 10 to 40 will be carried out for each enquiry)	
10	Redraw the Enquiry Access Path as a Jackson-style structure, lassoing all entity types linked by an Access Correspondence Arrow. This will now serve as the input structure.
20	Let the I/O Description be shown as an output data structure. If the function is an on-line enquiry, remove the boxes relating to input data elements. The structure left reflects the required format of the output report.
30	Identify the points of correspondence between the two structures, and merge them to form one process structure.
40	List the operations to be performed and allocate them to the new structure. The operations will be those to access records, navigate the Logical Data Structure, and perform semantic validation checks.
50	Design the formats for reports and inputs for the off-line parts of the enquiry functions. Much of this work may have been done previously with the User. Generally, the formats for the reports will reflect the output structure, and so there is little extra work to be done here, especially if the Application Style Guide has been completed.

↓

- Enquiry Processing Models
- Report Formats

Figure 6.7

Step 540   Assemble Logical Design

This step completes and allocates the Logical Design. It ensures the integrity and consistency of the respective products before passing them back to the project control organization.

Figure 6.8 illustrates Step 540.

- Dialogue Structures
- Dialogue Control Tables
- Entity Life Histories
- I/O Structures
- Function Definitions
- Menu and Command Structures
- Report Formats
- Required System LDM
- Enquiry Process Models
- Update Process Models
- Requirements Catalogue
- User Role/Function Matrix

↓

Task	Description
10	Examine the above products to ensure completeness and consistency. Amend the products if the reviews show this to be necessary.
20	Assemble the Logical System Specification, and publish it in accordance with the standards prevailing

↓

- Logical System Specification

Figure 6.8

Summary

Logical System Specification is the fourth Module in core SSADM. Its concerns are twofold:

1. To define the technical environment in which our new system must operate. This is achieved by selecting options for implementation in the form of hardware, software, development strategy, and configuration strategy (e.g., centralized or distributed). Many of these decisions will be dictated by the corporate IS strategy, and/or the findings of the Feasibility Study.

2. To develop the products from Stage 3, Definition of Requirements, into a logical specification for the system. This involves specifying logical processes for updating the database, and for making enquiries of the database.

   The techniques applied on this Module are:

- Technical System Options
- Dialogue Design
- Logical Database Process Design
- Entity/Event Analysis
- Function Definition

On completion of this Module, we are ready to move into Physical Design, in Module 5.

Figure 6.9 shows the Product Breakdown Structure for Logical System Specification.

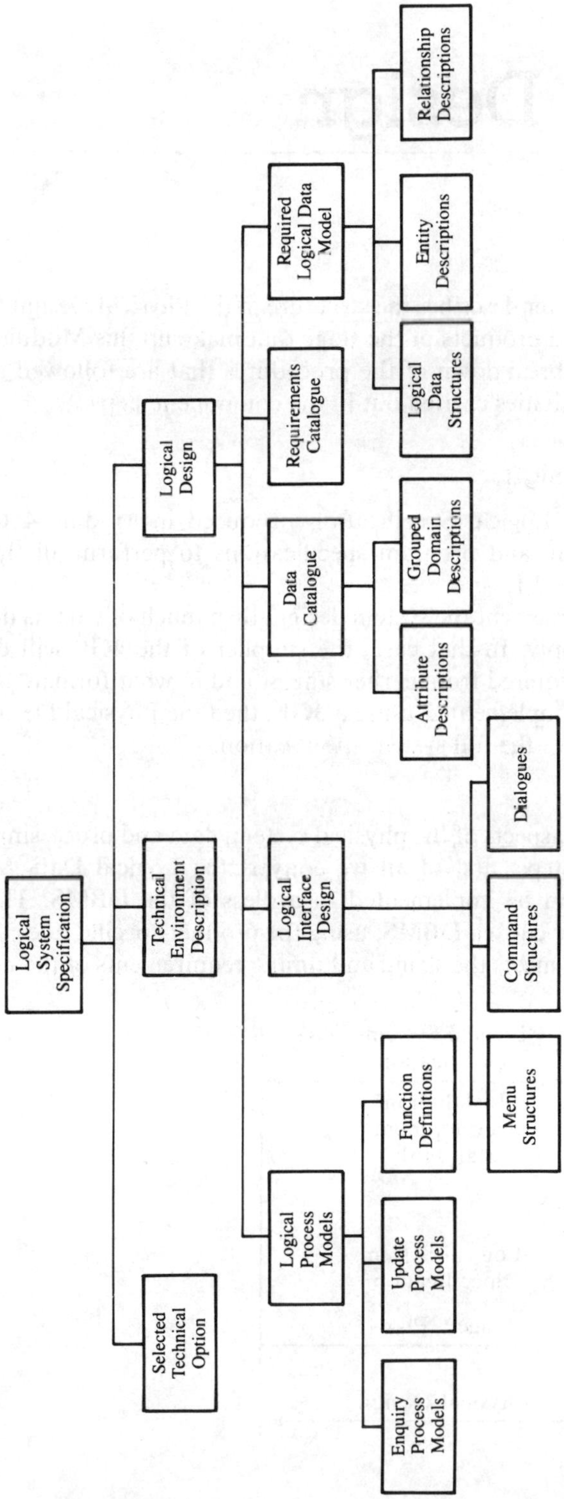

Figure 6.9

7. Physical Design

7.1 Aims of chapter

The first part of this chapter describes the structure of the Physical Design Module, as well as the objectives and products of the stage that make up this Module.

The second part is a breakdown of the procedures that are followed in Physical Design, detailing the activities carried out in the component steps.

7.2 Structure of Physical Design

This Module takes the Logical Specification produced in Module 4 to create a physical database design, and program specifications to perform all the required functions, as seen in Fig. 7.1.

If a 4GL is used to implement the system design, then much of what is described in this chapter will not apply. In that case, the supplier of the 4GL will detail what SSADM products are required from earlier stages, and in what format.

If the system is to be implemented using a 3GL, then the Physical Design Module will guide the designers to the full system specification.

Physical Design

This stage looks at both aspects of the physical system: data and processing. The data design is produced in steps: first of all we convert the Logical Data Model to a universal design that can be implemented regardless of the DBMS. This first-cut design is converted to the chosen DBMS, using the product-specific rules. The design is tuned to ensure that it meets the sizing and timing requirements of the new system.

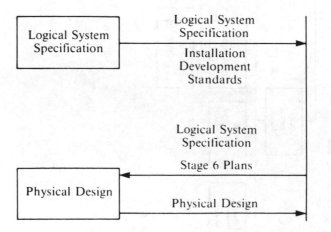

Figure 7.1

A strategy to implement this on the DBMS is drawn up, particularly where there cannot be a one-to-one mapping between the logical and physical data elements to be processed.

In conjunction with the data design, the Process Specifications are prepared according to the house style and format. A Process Data Interface is specified to enable the processes to carry out the DB accesses where the one-to one mapping is not possible, as indicated in the last paragraph.

STAGE 6: OBJECTIVES
1. To specify the physical data, processes, inputs, and outputs, using the language and features of the chosen physical environment, and incorporating installation standards.
2. The resulting design should provide everything needed to decide how application construction and introduction should be taken forward. User management will need to agree this with service providers in both systems construction and operations.

STAGE 6: PRODUCTS
All products from this stage are passed to data and control:
(a) Physical Data Design (optimized)
(b) Dialogue Designs
(c) Screen Designs
(d) FCIM
(e) Function Definitions
(f) Process Data Interface
(g) Required System Logical Data Model
(h) Requirements Catalogue
(i) Space Estimation
(j) Timing Estimation

Steps
610   Prepare for Physical Design
620   Create Physical Data Design
630   Create Function Component Implementation Map
640   Optimize Physical Data Design
650   Complete Function Specification
660   Consolidate Process Data Interface
670   Assemble Physical Design

7.3   Stage procedures

The seven steps listed above are not strictly sequential. Work on Step 610 can begin when the physical environment and implementation strategy have been chosen. All tasks must be completed before any subsequent steps are begun. Part of Step 620—that part which specifies database access components—must wait until Step 660 is complete, although the other tasks can be carried out after Step 610. Steps 650, 660, and 670 will be carried out in sequence.

Figure 7.2 illustrates the structure of the stage.

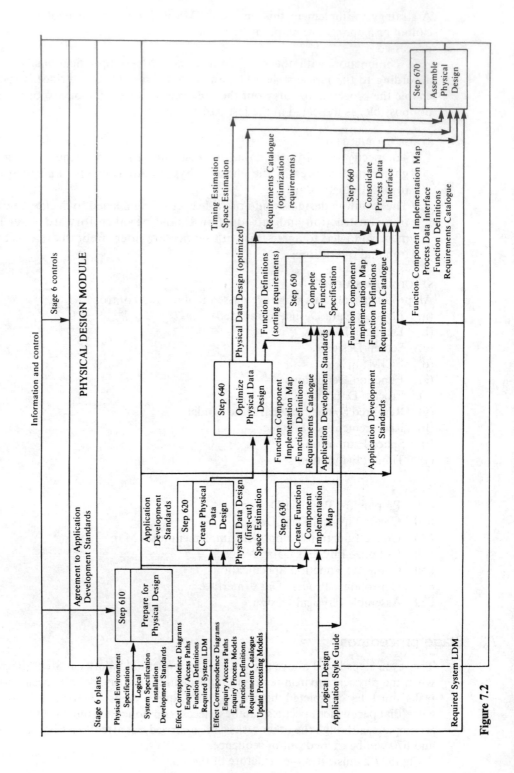

Figure 7.2

Physical Design

Step 610   Prepare for Physical Design

This step allows the design team to gain an understanding of the physical environment prior to carrying out Physical Design. A Physical Environment Classification scheme is used to categorize the physical environment. The scheme covers such things as data storage, performance, and processing characteristics. The characteristics and demands/constraints of the environment will clearly have an effect on the translation of the Logical Design. Decisions such as denormalizing, clustering, or indexing will be made at this time.

Also in this step, the Application Development Standards are defined. The three main tasks here are: to define the standards for the use of the physical processing system; to define the Program Specification formats; and to develop the Activity Descriptions for the Physical Design activities that are specific to the implementation environment It may be that specification formats already have an installation standard, in which case, of course, nothing is to be done on that task.

The techniques employed are:
- Physical Data Design
- Physical Process Specification

Figure 7.3 illustrates Step 610.

- Installation Development Standards
- Logical System Specification
- Physical Environment Specification

↓

Task	Description
10	Complete the Processing Systems Classification for the processing environment. The vendors of the DBMS should supply the data with the package, but this is not an invariable rule. If the storage and performance details are not supplied, the designer must set about finding the data, probably by creating two forms that can be easily filled in. Complete the DBMS Performance Classification. Complete the DBMS Storage Classification.
20	Design forms for the DBMS Space and Timings Estimations.
30	Specify the standards for the use of the physical processing system and the DBMS facilities. These may often be inferred from other systems running at the installation.
40	Specify the product-specific data design rules, unless already available.
50	Specify the naming standards for the application.
60	Develop the Program Specification and data design standards by defining a Product Breakdown Structure and Product Description for the Physical Design. Develop an Activity Network and Activity Descriptions for the remainder of Stage 6.
70	Initiate the preparation of manuals—user, operations, and training. The production of these manuals will continue into the construction phases of the project, after SSADM is complete.
80	Agree the Physical Design Strategy with the project board

↓

- Application Development Standards

Figure 7.3

Step 620   Create Physical Data Design

The aim of this step is to develop a Physical Data Design to implement the Required System LDM. An algorithm to convert the LDM into a universal first-cut design is followed. This follows general principles, and the first-cut design is non-specific.

This first cut is then converted into a design specific to the DBMS to be used, using its own rules.

Techniques employed in the step are:

- Physical Data Design

Figure 7.4 illustrates Step 620.

- Application Development Standards
- Effect Correspondence Diagrams
- Enquiry Access Paths
- Function Definitions
- Required System LDM

↓

Task	Description
10	Identify the features of the LDM that are required for Physical Data Design.
20	Identify the required entry points and distinguish those that are non-key.
30	Identify the roots of physical hierarchies.
40	Identify the allowable physical groups for each non-root entry.
50	Apply the least dependent occurrence rule.
60	Determine the block size to be used.
70	Split the physical groups to fit the required block size.
80	Apply the product-specific data design rules for the target DBMS.

↓

- Physical Data Design (first cut)
- Space Estimation

Figure 7.4

Step 630   Create Function Component Implementation Map

This step is to specify those functions that are not included in the Logical Design, and also to describe those function components that can be specified non-procedurally.

The components not defined in the Logical Specification (syntax error handling, physical I/O formats, physical dialogues, etc.) are specified. The Function Component Implementation Map (FCIM) identifies duplicate and common function components, and defines the relationship between all the function components.

Specification of database access components is not carried out until Step 660, as a part of the Process Data Interface.

Figure 7.5 illustrates Step 630.

- Application Development Standards
- Logical Design

  ↓

Task	Description
10	Identify and remove duplicate processing.
20	Identify and remove the specification of common processing.
30	Define success units for each enquiry and update process.
40	Specify syntax error handling.
50	Specify controls and error handling.
60	Specify physical I/O formats.
70	Specify physical dialogue design.
80	Specify those function components that can be described non-procedurally.

↓

- Function Component Implementation Map
- Function Definitions
- Requirements Catalogue

Note: tasks 30–80 are performed for each function.

Figure 7.5

Step 640   Optimize Physical Data Design

This step is to tune the first-cut product-specific data design to meet the space and timing objectives.

The Physical Design is validated against the performance levels defined in the Function Definitions and Requirements Catalogues. If the preset performance levels cannot be met, then the data design will be tuned.

The technique employed is:

- Physical Data Design

Figure 7.6 illustrates Step 640.

- Application Development Standards
- Effect Correspondence Diagrams
- Enquiry Access Paths
- Enquiry Process Models
- Function Definitions
- First-cut Physical Data Design
- Requirements Catalogue
- Space Estimation
- Update Process Models

↓

Task	Description
10	Estimate the storage requirements. If necessary, restructure the data design to meet the requirements. It is desirable to preserve a one-to-one logical-to-physical mapping if this is feasible. In many cases, of course, it will not be possible or practicable to do this.
20	Estimate the resource times of major functions. If the performance levels of these critical functions are not acceptable, alter the structure to achieve the desired access in time. Exploit the access mechanisms of the target DBMS to improve speed of access, but preserve the one-to-one mapping of logical-to-physical where possible.

↓

- Function Definitions
- Optimized Physical Data Design
- Requirements Catalogue
- Space Estimation
- Timing Estimation

Figure 7.6

Step 650   Complete Function Specification
In this step we specify those components of a function that cannot be specified procedurally. The results will be in sufficient detail for a programmer to code them, or for an application generator to produce the code.

This step is not an invariable step: it is only undertaken if components of the FCIM are to be specified non-procedurally. Where there are structure clashes, Specific Function Models will be produced. All function components needing procedural code will have program specifications written for them.

The technique employed in this step is:

- Physical Process Specification

Figure 7.7 illustrates Step 650.

- Application Development Standards
- FCIM
- Function Definitions (sorting requirements)
- Logical Design
- Requirements Catalogue

↓

Task	Description
10	Distinguish the logical processes in the function. If there is not sufficient detail in the Function Definition, draw a Specific Function Model (SFM). This will be necessary in the case of, for example, a structure clash in the logical processes which has not yet been resolved. The SFM will define extra Modules, such as sort processes, that do not appear on the Universal Function Model (UFM). If you do need to produce an SFM at this point, all the syntax error handling and I/O routines must be specified on them.
20	Combine the logical processes into a physical Program Specification or success unit.

↓

- FCIM
- Function Definitions
- Requirements Catalogue

Figure 7.7

Step 660   Consolidate Process Data Interface
In this step we complete the procedural specification and examine the non-procedural implementation of the mapping between the Physical Design and the logical view of the data.

We compare the FCIM data access components with the optimized data design to identify any mismatches in the views of the data. Any such mismatches are resolved by identifying the keys and then the navigational sequence needed to pass the required view to the FCIM.

This may be implemented by a coded Module, procedurally, or with a non-procedural language, such as SQL, if this is used at the developer's site.

The set of Process Data Interface (PDI) components is rationalized, and any special maintenance or enhancements requirements recorded. Design compromises are recorded in the Requirements Catalogue against the affected requirements.

The technique employed in this step is:
- Physical Process Specification.

Figure 7.8 illustrates Step 660.

- Application Development Standards
- FCIM
- Function Definitions
- Optimized Physical Data Design
- Required System LDM
- Requirements Catalogue

↓

Task	Description
10	Identify mismatches between a data access function component and the optimized data design. Step 640 will have identified some of these.
20	For each such mismatch, identify the physical keys of the master and details to be accessed.
30	Starting at the top of each hierarchy affected, determine the physical access sequences to provide the FCIM access components. This will be done either with a procedural specification, or with the non-procedural language already identified.
40	Compare all the new processing components to identify duplicates.
50	Within the FCIM, record all the PDI elements that handle mismatches as being the subject of special maintenance and enhancement requirements. Note any which use special features in the physical environment, and those where low-level routines may be appropriate for performance reasons.
60	Annotate the Requirements Catalogue to show design decisions which limit the extent to which requirements have been met.

↓

- FCIM
- Function Definitions
- PDI
- Requirements Catalogue

Figure 7.8

Step 670   Assemble Physical Design
The Physical Design Module is complete at the end of this step, and the final Physical Design, the end of SSADM involvement in the project, is assembled and delivered.

Each product has its own quality criteria that have been built into the Product Description, and now we carry out a check for consistency between all these products.

Figure 7.9 illustrates Step 670.

- FCIM
- Function Definitions
- Optimized Physical Data Design
- PDI
- Required System LDM
- Requirements Catalogue
- Space Estimation
- Timing Estimation

↓

Task	Description
10	Check the completeness and consistency of the Physical Design products by reviewing the inputs to this step, listed above.
20	Assemble and publish the Physical Design.

↓

- Physical Design

Figure 7.9

Summary

In this Module we have translated the Logical System Specification into a Physical Data Design and Program Specifications. The actual working system should now be ready to be written by a programmer using 3GL techniques, or to be fed into an application generator to produce the code.

Emphasis is placed on standards that are not supplied by SSADM, but which belong to the local installation, or are specific to the vendor's own product, be it DBMS, 4GL, application generator, etc.

Much of the advice and guidance in Physical Design Module is necessarily generic, and will be adapted to whatever physical implementation vehicle is selected. As always, it is up to the design team and Project Management to apply the guidance intelligently and appropriately.

Figure 7.10 shows the Product Breakdown Structure for the Physical Design Module.

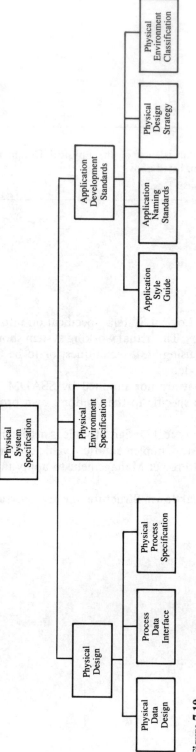

Figure 7.10

PART TWO SSADM techniques

Case study

The following case study will be used to demonstrate the techniques of SSADM throughout Part 2. The scenarios given form the basis for Data Flow Modelling, Logical Data Modelling, Entity/Event Modelling, Relational Data Analysis and Logical Database Process Design.

As this study is intended only to demonstrate the different techniques, rather than act as an example of the whole method, I shall not produce complete solutions for each one, just sufficient to illustrate how it is used.

Super Systems plc

Super Systems (SS) plc is a nationwide communications organization employing 5000 staff, and is based at 70 centres across the country. The company carries out its own staff training at the chief training centre (CTC) in Milton Keynes, and at 10 other regional training centres (RTCs). The training requirements are wide, embracing clerical and management skills, technical engineering skills, electrical engineering skills, warehousing and stores skills, computing and software engineering skills.

The CTC is responsible for scheduling and administering all courses across the company. The Administration section at the CTC is split into Course Maintenance (CM), Course Scheduling (CS), Bookings (BK), and Billing (BI). Preliminary interviews with these functions revealed the following tasks and activities.

A   COURSE MAINTENANCE (CM)

This section is responsible for publishing the internal brochure. The Course Managers pass information about their courses to CM. This information is related to:
1. The creation of a new course title, with information about objectives, prerequisites, duration, subjects covered, maximum number of students, and level of staff who should attend (e.g., clerical staff, senior storekeeper, line manager).
2. Amendment to published details—should any of the items in 1 above alter.
3. The removal from the portfolio of a discontinued course.

B   COURSE SCHEDULING (CS)

CS publishes details for each run of the course titles, giving the location, start and end dates, and the name of the course contact (this will be either the Bookings officer, or the Course Director).

A course that is in common demand will be scheduled to run on a regular basis: monthly, fortnightly, weekly, etc. Courses that have a limited take-up will be scheduled only when there are enough names on the waiting list to fill a run.

Each week CS examines the waiting lists for each course title; if there are sufficient to run the course, CS then examines the Training Schedule for free dates, and books a run of the course between three and six weeks hence. Facts that they need to establish are availability of premises and availability of appropriate teaching staff. A training location (usually an RTC) will be booked at the same time as the course is scheduled; staff will be allocated if they can be identified at the time, but they may not be allocated until later.

When such a course is scheduled, the dates and delegate names are posted to the Course Manager (who will appoint a Course Director), the Nominating Manager of the delegates, teaching staff and Bookings.

No two runs of the same course will take place concurrently.

C   BOOKINGS (BK)

BK receives queries and requests for places from managers or Training Officers from SS's branches. If the request is for a course title with no scheduled runs, the names will be placed on a waiting list; if there is a run scheduled, the delegate's name will be entered for that if there is still space. If the course is full, the delegate's name will be put on the waiting list for the next scheduled run.

BK receives details of all courses scheduled from CS when the dates are agreed. Seven working days before each scheduled course begins, BK sends joining instructions to each delegate's Nominating Manager or Training Officer, whoever is nominated to receive the instructions.

If a delegate cancels before a course, BK tries to fill that place with someone from the same unit. If this is possible, the names are changed, and no further action is taken. If not, the next name from the waiting list is substituted. If the delegate cancels less than two weeks before the course is due to begin, the transfer fee is forfeit. Otherwise it will be refunded, or credited for another delegate from the same Nominating Manager.

If for any reason a Course Manager or CS cancels a course, BK is notified and has the task of informing all delegates' nominators.

A list of all attendees who have been booked for scheduled courses is sent to BI at the end of each week. Arrangements for transfer fees can then be made.

At the end of a course, details of attendance and performance are kept against each delegate's record, in order to provide information for annual appraisal, provide information on a delegate's training path, and also investigate subsequent query or complaint.

D   BILLING (BI)

BI is responsible for sending requests for payment for each course run. Each booking is made by a Nominating Manager for one or more delegates. Payment (transfer fees, not cash) is made each month, and is for all the manager's delegates booked on to a course.

An arrangement may be made by which a special booking will be billed separately from the regular monthly bill, and paid at a special rate.

BI is responsible for both maintaining the price book for courses—a loose-leaf binder listing all course rates—and notifying all RTCs and Training Officers of any changes. They receive from Management—and distribute across the company—an amendment list of new prices twice a year.

Use of case study

For each of the exercises involving SS plc, appropriate levels of detail will be given, if necessary, to enhance this outline. In this overview, for example, little mention is made of problems or requirements. Those chapters dealing with Requirements Definition and Function Definition will identify some of the problems and requirements that pertain to this system.

There are some instances in the text where SS plc does not provide an example of an aspect of a particular technique under discussion. When this happens I will draw on other examples as appropriate, particularly an employment agency called Jobs-For-U (JFU). This case does not need the same level of detailed explanation.

Job-Seekers approach JFU, who hold information on vacancies supplied by Employers. The vacancies require skills, belonging to particular skill areas, such as shop work, skilled machinist, purchase ledger maintenance, HGV driver, etc. JFU marry up the Job-Seekers with Vacancies, according to these skills and qualifications, and arrange interviews. The outcome of an Interview may be a rejection, or a job-offer, which in turn may be accepted or rejected.

8. Logical Data Modelling

8.1 Aims of chapter

In this chapter you will learn:
- The use of Logical Data Modelling in SSADM
- How to identify entities
- How to chart and label relationships
- How to show optionality in relationships
- How to model a bill-of-materials explosion
- How to model entity subtypes
- How to cross-validate the DFD with the LDS
- How to validate Logical Data Structures with Enquiry Access Paths
- How to document Logical Data Structures

8.2 Where LDM is used in SSADM

The LDM is created early in the SSADM cycle, and updated at several points subsequently.

Step 110   Produce overview LDS.

Step 140   Produce an outline LDS of the data in the current system and identify attributes.

Step 150   Convert DFD of current services to logical view. Use LDM to define Logical Data Stores.

Step 210   Prepare BSOs, and support with LDM.

Step 310   Use LDM to update low-level DFDs with new data items.

Step 320   Extend current model to support new processing requirements. Fully document entities and attributes.

Step 340   Carry out RDA on I/O Descriptions, and use the results to validate and enhance the LDM.

Step 360   Use the LDS to create the Entity Life Histories in two passes—bottom-up first, and top-down for showing abnormalities. The LDS may be extended or amended in the light of this activity.

Step 370   Review the Requirements Catalogue for both functional and non-functional requirements. Amend the Required LDM as appropriate.

Step 520   Complete the process definitions for each update function, using the ELHs, Required LDM and Effect Correspondence Diagrams (see Chapter 11).

Step 530   Process Definition for all enquiries will be completed here, with enquiry aspects of Update Processes. Inputs will be the Required LDM and the Enquiry Access Paths.

Step 620   The Required LDM is converted into a format which will then be implemented on a product-specific DBMS design.

LDM in SSADM

Inputs to LDM are as follows:
- Feasibility Report
- Project Initiation Document
- Overview LDS
- Requirements Catalogue
- Selected Business System Option

Outputs from LDM are as follows:
- Logical Data Structure
- Attribute Description
- Entity Description
- Domain Description
- Relationship Description

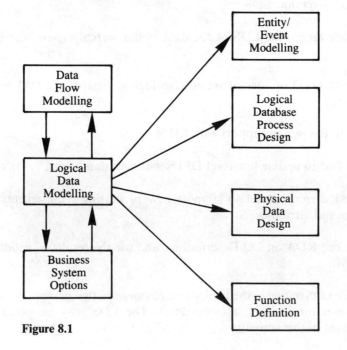

Figure 8.1

Figure 8.1 illustrates the relationship between LDM and other SSADM techniques.

8.3   Logical Data Structures

Most descriptions of the current system, such as narrative, or Data Flow Diagrams (see Chapter 9) show us how data is stored in the workings of the system, but this is usually in a very unstructured manner. By 'unstructured', I mean that *ad hoc* storage of data items, especially in a manual office, puts clusters of data items together for convenience rather than logic.

Let us take details of one of SS plc's courses for example. If we look at the form that is used to make booking (Fig. 8.2), we see that we have information of Courses, Scheduled Courses, Locations, Managers, and Delegates, all on the same piece of paper.

How could we store this as a computer file? What does it record? Manager? Delegate? Course? Scheduled Course? It covers all of those, and more. Any record that is to be stored, though, must deal with a single topic or classification, not half a dozen different ones.

Logical Data Modelling is a technique for examining the *structure* of the data that is hidden inside such documents. It is logical because it ignores the physical manner of the storage in our system.

The technique is, like most of SSADM, based on a diagrammatic notation for ease of creation, amendment, and understanding. The notation consists of just two components, representing *entities* and *relationships*.

Entities

By 'entity' we mean something in our system about which we need to hold information. On our booking request form (Fig. 8.2), we asked: What is the subject? Manager? Delegate? Course? The fact that we can ask such questions points each of those out as candidate entities.

An entity must have the following characteristics:
- It must contain information of interest to the system.
- There must be the possibility of more than one occurrence—the environment itself (SS plc) cannot be an entity as there is only one of it.
- Each occurrence must be uniquely identifiable—there must be a code or key for each entity.

Generally, when classifying data into entity types, we think at first in terms of physical 'things', such as a Person, or a Stock Item. Data is more complex than that, and can be classified in a wider variety of ways. Different classifications of entities include:
- Tangible, e.g. Person, Product, Car
- Conceptual, e.g. Course Title, Account, Job
- Active, e.g. Delivery, Payment, Sale
- Permanent, e.g. Warehouse
- Volatile, e.g. Stock

DEPICTION
An entity is represented by a 'soft' box (i.e. one with rounded corners), as in Fig. 8.3.

Booking Form

Super Systems : Training

SST Course Booking Duty:	

Course Title

Delegate Details (Please advise of special requirements/disability):

Surname

First Name

Division ☐   Section ☐   Group ☐

Nominating Manager:

Duty

Telephone

Branch

Signature	Date

Figure 8.2

Figure 8.3

IDENTIFICATION

Each occurrence of an entity must be uniquely identifiable: this implies the use of a code. When carrying out your investigation of the environment, look for any code numbers in use. If the code is able to identify uniquely such an occurrence, you have found an entity. Common codes in use are such things as Personnel Number, Part Number, Account Number, Invoice Number, etc.

Many codes will, in fact, be combinations of other codes. If this is the case, check to see if each part is unique or common. Take the example of a company handling air freight. Let us postulate a document, Airbill, which identifies a consignment for a particular airline, stored in a particular shed. The code which identifies each one is made up of: Airline Code, Shed Code, and Bill Code.

If Bill Code is unique within the system, then we have found *four* entities: Airbill, Airline, Shed and Bill.

If, on the other hand, Bill Code is repeated for each airline, then we have still found four entities: Airbill, Airline/Bill Airline, and Shed. In this case Airline/Bill has a *composite key*, i.e. one made up of two or more items, one of which is not unique in the system, and at least one of which is unique in the system. This question of key types is expanded in Chapter 11.

When identifying entities by keys, therefore, look for the simplest identifier for a group of data items. That may be one data item alone, as in Airline, or it may be a composite, as in Airline Bill. In both cases, the identifier is the simplest possible.

Attributes

Every entity contains a set of data items, or *attributes*, which both identify and describe the entity. Thus, if the entity were Product, possible attributes to describe it might be Description, Price, Colour, Bin Location. All the attributes for an entity describe the extent of our interest in that entity. Thus, if our interest in Product included details of who supplies it, and how much was in stock, we would need two extra attributes, Supplier and Quantity Held.

Domain

A *domain* is a permitted range of values for an attribute. Date of Birth, for example will have a domain that is different for particular entities: a baby in a clinic will have a domain that goes back no further than, say, three years; a student at a college, on the other hand, will have a domain for Date of Birth that will be no more recent than the previous sixteen years.

A serial code of five numbers will have a domain of any possible combination of five numeric characters, or to put it more formally, 00001–99999. If the code in the system covers only the range 01000–50000 for some reason, then that range is the domain for that code.

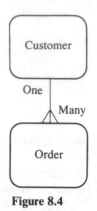

Figure 8.4

Relationships

A *relationship* is a logical business association between two and only two entities. In a mail-order business, for example, a customer will place an order for goods or Catalogue Items. Thus the three entities, Customer, Order and Catalogue Item, are related to each other. The relationships are shown by a straight line linking the entities.

Relationship in SSADM are normally in the form known as one-to-many, i.e., in a relationship one entity will occur once only, and the other may occur many times (see 'Cardinality', below). The relationship joins the 'many' entity in what is commonly known as a 'crow's foot'.

Figure 8.4 illustrates the CUSTOMER–ORDER relationship, showing the one-to-many relations.

Let us identify two entities from SS plc:

- Course title, with a key of the first three letters of the course title (Course Code), is an entity that describes the titles in the brochure.
- Course Run, with a key of Course Code and Start Date, is each occasion that that course title is presented to a class of students.

There is a necessary relationship between these two entities in that Course Run refers directly to a specific Course Title, and Course Title is given several presentations, or runs, to classes. Each relationship has several features that must be considered. They are:

- Cardinality or degree
- Optionality
- Names

CARDINALITY (DEGREE)

This term simply means one of the following facts about a pair of entities: one occurrence of an entity in the relationship is associated with just one occurrence of the other $(1 : 1)$, or each occurrence of one entity in the relationship can be associated with more than one occurrence of the other $(1 : m)$. The entity at the '1' end of the relationship is termed the *master* and the entity at the 'm' end is termed the *detail*. Finally, occurrences of either entity in the relationship may be associated with more than one occurrence of the other $(m : n)$.

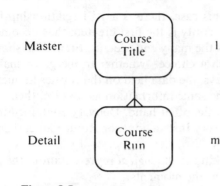

Master

Course
Title

1

Detail

Course
Run

m

Figure 8.5

That end of a relationship line that is attached to a detail entity has the crow's foot attached, to denote 'many'. Figure 8.5 tells us that one Course Title will have many Course Runs, while each run of a course will be for one and only one Course Title.

Permitted cardinality

Traditionally, data modelling in SSADM has accepted the existence only of $1 : m$ relationships. However, in the description of the Current Environment, one-to-one and many-to-many relationships may exist. These will be recorded as such in the description of Current Services, but will be resolved into one-to-many before the required LDS is complete.

1 : 1 relationships   In many cases, $1 : 1$ relationships are in fact a description of just one entity: the other entity merely contains extra information rather than information about a different topic. When this is the case, the entities should then be merged to form one.

If this is not the case, the relationship should be shown as $1 : m$ instead. One of the entities in the relationship will have been formed first, so make that the single occurrence, and the one created second becomes the many end with the crow's foot.

An example of this is the situation where a company invoices customers for goods delivered rather than goods ordered, in other words, they present the bill on delivery,

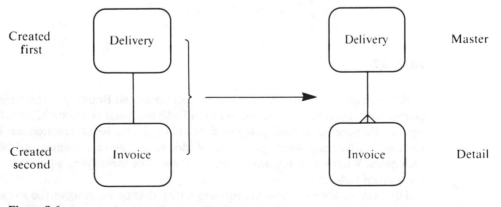

Created
first

Delivery

Delivery

Master

Created
second

Invoice

Invoice

Detail

Figure 8.6

even if it is only a part delivery. In this case, there is a 1 : 1 relationship between Delivery and Invoice. As the Delivery entity is the first created, that becomes the 1 end (master), and the Invoice becomes the many end (detail). Figure 8.6 shows this.

When the analyst is faced with such a choice—whether to merge or make 1 : m the m most important factor is, as always, the data held on the entities in question. If in the example above, Delivery held the same information as Invoice, then to merge would be the optimum solution. If, on the other hand, Delivery contained information irrelevant to Invoice, such as Delivery Instructions, or Route Code, there might be a case for separating out the entities.

Remember, at this point we are looking at the *logical* representation; the *physical* grouping does not affect such choices at the moments.

m : n relationships   If you uncover an *m : n* relationship, it may be that there is another entity that you have not yet recognized. In SS plc, we find that a branch makes a number of bookings for a number of our courses. Each Booking may be for a number of Delegates, but for one Course only. Each Delegate may be booked on to a number of courses, i.e., be the subject of a number of Bookings.

At first look, we identify the following entities:

- Course
- Booking
- Branch
- Delegate

This is modelled in Fig. 8.7

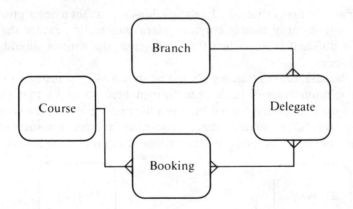

Figure 8.7

We have an *m : n* relationship between Delegates and Bookings, which may need to be resolved. There is information on each Delegate, such as Name, Contact Address, Special Requirements and Telephone Number. There is information on Bookings, such as Authorizing Manager, Date of Booking, Contact Name, and Number of Delegates. What we do *not* have, though, is a way of identifying a particular Booking for a given Delegate.

This tells us about a possible missing entity that could resolve the *m : n* relationship. The entity acts as a link between Booking and Delegate, and for its key will take

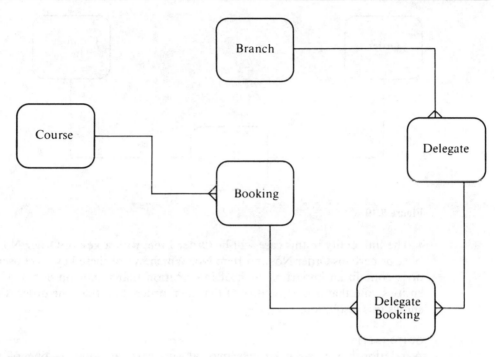

Figure 8.8

a combination of Booking No. and Delegate No. Naming the entity sometimes causes a problem, but in this case Delegate Booking will clearly indicate the nature of the entity.

When a *link entity* is identified and inserted into the diagram, it becomes the detail of the two entities that it reconciles, as illustrated in Fig. 8.8.

Sometimes the link entity will contain additional information in the form of data items (attributes) that are not present on either of the two original entities, and sometimes the link entity itself is all the extra information held.

In our current example, Delegate Booking tells us nothing new about the Delegate or the Booking *per se*. Its extra information value lies in identifying the one Booking out of many for a particular Delegate. This would be important both for charging purposes and for monitoring the individual's development plan.

In a different environment, now, if we look at an Order Form in the Mail Order Processing system, we can identify an *m* : *n* relationship between Order and Catalogue Item, in that each Order Form will be for many items, and the Catalogue Item will be the subject of many Order forms, as shown in Fig. 8.9.

Figure 8.9

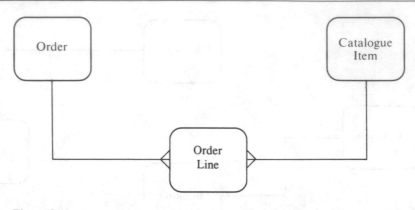

Figure 8.10

The link entity in this case will be Order Line, with a key of Order No. and Line No., or perhaps Order No. and Item No. Whichever of these keys is chosen, the link does contain an important item of information that is not on either of the other entities, and that is the quantity of that item ordered on that one order. Figure 8.10 shows this resolution.

OPTIONALITY

A relationship between entities may always exist as soon as one of the entity occurrences comes into existence. Figure 8.11 shows a Booking for many Delegates for a Course Run. As soon as a Booking is created, there is a necessary link between Booking and Course, and Booking and Delegate. It cannot exist without at least one of each being attached to it. This means that the relationship between Booking and the other two is *mandatory*.

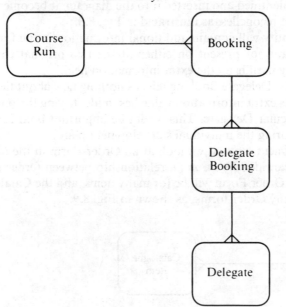

Figure 8.11

Other entity occurrences in a relationship may exist even if there are no occurrences of the entity in existence at the other end. Using the same example, a course can be scheduled to run, and so the occurrence of Course Run exists before any bookings have been made. In this case, the relationship between Course Run and Booking is said to be *optional*.

A mandatory relationship is depicted by a solid line leaving the entity. An optional relationship is depicted by a broken line leaving the entity.

NAMES

One way of introducing rigour into LDM is by forcing the analyst to name each end of the relationships. It is not very useful to say simply, 'Each occurrence of entity A has one or more occurrences of entity B attached to it.'

To show that we have fully understood the nature of the entities and their relationship, both master and detail ends should be described in a succinct manner. Find a meaningful word or phrase that connects them, rather than the weak 'has', and use the following formula:

- Each occurrence of entity A must/may (link word) with/to/for ... one or more occurrences of entity B.

for one end, and, for the other end:

- Each occurrence of entity B must/may (link word) with/to/for one and only one occurrence of entity A.

The selections 'must' or 'may' in each of these statements reflects whether that relationship end is optional or mandatory. If the relationship is mandatory use the word 'must', and if optional, use the word 'may'.

Figure 8.12

Figure 8.12 shows the relationship between Course Run and Booking. The relationships are named as follows:

- Each Course Run *may* have one or more Bookings against it.
- Each Booking *must* be for one and only one Course Run.

Relationships can be mandatory or optional, as we have seen; but that is not the whole story. As we explore the environments more deeply, we may uncover such relationships as *exclusive* and *recursive* relationships, the latter finding its greatest expression in the structure known as *Bill of Materials Processing* (BOMP) (explained in more detail later).

Exclusive relationships

Figure 8.13 shows an exclusive relationship between three entities in a motorway patrol system. The entities are: Breakdown, Patrol Vehicle, Salvage Contract.

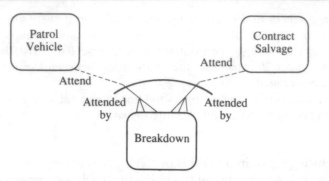

Figure 8.13

The diagram shows that a Breakdown may be attended by a Patrol Vehicle or a Contract Salvage vehicle, but not both. The arc is known as an *exclusion arc*, and denotes the either/or relationships. In this case, the exclusion works from one detail type to one of a number of possible master entity types.

The exclusion can work in the opposite direction, with one master being attached to one and only one of two possible detail entities. An example would be a train that could be attached to passenger carriages, or to goods wagons, but never to both. In the diagram, the master could have any number of different entity types as details, but any one occurrence of the master could be connected to only one detail type. This is illustrated in Fig. 8.14.

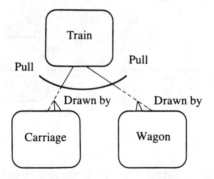

Figure 8.14

Exclusion was described earlier as indicating membership of one and only one set. This is simple, but not strictly accurate. There may be a number of different detail types belonging to one master type, yet a single occurrence of the master may own more than one detail type. If this should prove to be the case, the exclusion arcs should be numbered or labelled in some way to show the business rules that govern membership.

To illustrate this, Fig. 8.15 shows a sports federation, whose Teams may be made up of either all Professional or all Amateur players, who may be either all Male or all Female. Using the notation in the diagram, we can ensure that there is no mix of professional and amateur on the one hand, or male and female players on the other.

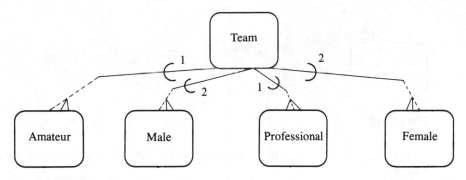

Figure 8.15

Bill-of-Materials

Some occurrences of an entity type may be related to other occurrences of the same entity type. The manufacturing Bill of Materials Processing (BOMP) structure is the classic example of that. Figure 8.16 depicts BOMP.

Each entity on the structure has the same key—Part No. The link entities will have a compound key of Part No./Part No., each Part No. reflecting one of the master keys.

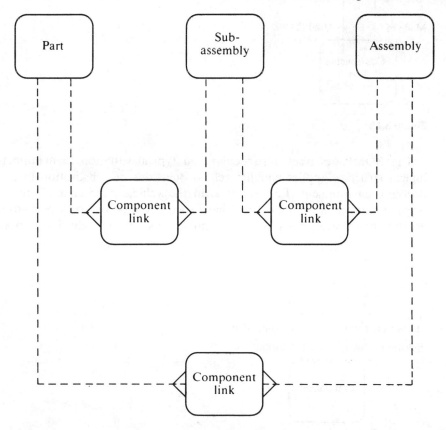

Figure 8.16

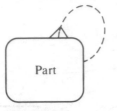

Figure 8.17

The diagram can therefore be drawn more simply as a recursive structure, as Fig. 8.17.

If the relationship is many-to-many, that is resolved by inserting a link entity, with a double relationship, as in Fig. 8.18.

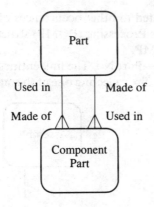

Figure 8.18

The BOMP describes a particular and typical situation in manufacture. This logical structure applies equally well to questions of substitution in a supply or ordering environment. For example, in the vehicle maintenance unit of a large organization, each car model will have specifications for every part, from engine, gearbox or radiator, down to bushes, nuts, bolts, washers, etc. Data on all of these

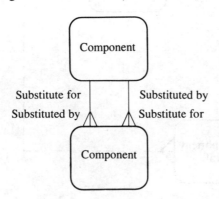

Figure 8.19

will be stored under the key of Component No. If the component in question is unavailable when a vehicle is in for repair or service, some other specification may be an acceptable substitute. This will be represented in our model according to Fig. 8.19.

Entity subtypes

A common problem when drawing up data models is how to resolve a situation where two or more entities have been identified which share properties and behaviour, yet also have significant differences.

An example can be found in a college, where employees may be Academic staff, with a particular grade, history, specialism, etc., or Ancillary staff, with a completely different contract, terms and conditions, and responsibility: both share some attributes, such as Employee No. and Pay Code, yet have different information requirements apart from these. How is this to be represented?

We could have just one entity Employee, with far more attributes available than any one occurrence will require, or we could employ the concept of *subtyping*.

We would draw an entity *supertype*, employee, with two *subtypes*: Academic and Ancillary. Both types of Employee may require sick leave, and so records of Sickness would be attached to the Employee supertype. Only Academic staff, however, would need records kept of research projects, or publications, so an entity Publication would be attached only to the subtype Academic.

There are different possible ways of recording this on an LDS. The *SSADM Manual* recommends a 1 : 1 exclusive relationship between supertypes and subtypes. Using that notation, the college staff model would be as in Fig. 8.20.

Another common notation, which is perhaps easier to recognize in a busy diagram, is to incorporate the subtype boxes inside the supertype, as in Fig. 8.21.

The question of notation is not as crucial as many students—or practitioners— seem to think. Within a project, and preferably within an institution, the notation should be agreed and published as a standard. Which notation should be adopted is less important than the thought and analysis which go into resolving the situation.

Subtyping is a more common occurrence in the real world than many introductory texts imply, and so it is worth pointing out certain features that the practitioner is sure to encounter, and may have to resolve.

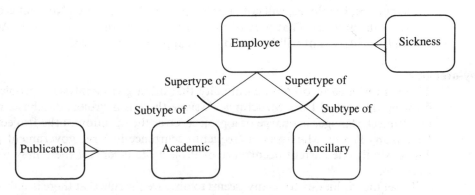

Figure 8.20

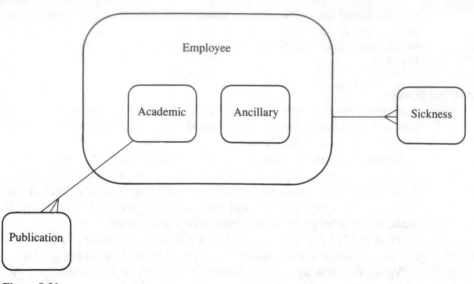

Figure 8.21

- Each subtype 'inherits' characteristics from all its supertypes. This means that attributes common to each subtype will be held only in the supertype. Thus, Employee would hold details of name, address, date of birth, next of kin, and so on, which would apply to both subtypes. Information which applied only to the academic staff, such as Lecturing Grade, would be held only on the Academic subtype.

  There is no reason why there should not be nests of subtypes, or even networks of supertypes and subtypes, if these provide accurate information about the business rules in our particular sphere of interest.
- Each level of subtyping will introduce further data items, relating only to that subtype.
- As an occurrence of a subtype chain represents one thing only, each occurrence of an entity type can be in only one subtype at a given level.
- Each supertype must contain a *classifying attribute* which acts as the unique identifier of the subtype. This means that in the college employee example, the supertype, Employee, will contain that part of the Employee No. that is common to both subtypes. That part of the key which differentiates between Academic and Ancillary staff will be found in the appropriate subtype.

Many-over-time

There is one area to do with cardinality that often gives analysts a problem when drawing up Logical Data Structures. Assume there is a Project database, with just one Project Manager in charge of a project. Over the duration of the Project, it may have many Project Managers, a frequent occurrence in large government projects. There is only one current occurrence of Project Manager, yet there may be 'many-over-time'.

To set the cardinality to 'many' seems to obscure the rule that there is only ever one at a time; to set it to 'one', however, ignores an important rule about data on previous managers needing to be retained.

The answer must be, first of all, to be aware that such a situation exists, and then to mark it on the supporting documentation. As long as it is recorded, it does not matter so much whether it is marked as one or many. However, it *must* be recorded on the Relationship Description, as it will have an impact on storage requirements, and on meeting the User's data requirements.

In another example, it may be that a company needs to keep information about its vehicles, including their Service History. In such a case, the analyst must identify precisely what information is required. If the User wants full information on each service, then an entity—Service History—will be needed. If all that is required is date and mileage of last service, then appropriate attributes on the vehicle entity are sufficient. The factor is, how much information needs to be held over time: current, or all history?

8.4   How to derive the LDM

While there is no rigid algorithm for creating a Logical Data Structure, the following guidelines will help beginners, as well as providing experienced practitioners with a checklist of activities.

Identify the candidate entities

During the fact-finding exercise at the beginning of analysis in Step 110, identify as many codes as you can that are used in the system, or will be used if this is a greenfield system.

Having listed the candidate entities, criticize the list with the Users. Be careful to identify synonyms (two names for the same logical entity, perhaps by separate departments) and homonyms (one name for two separate logical entities).

Once you have identified entities, be sure that the Users are happy with the name chosen. The entity will not reflect the way that the data is held in the physical system. For example, an order processing system may hold data stores for Received Orders, Pending Orders, Partially-filled Orders and Filled Orders. These data stores may be scattered across two or three rooms. All, however, refer to the business entity Order. Do not create separate entities for Pending Order and Filled Order: identify the underlying entity Order and let its attributes summarize all the different data stores relating to it.

Further entities may be identified later, during Step 340, for instance, using RDA, and during Step 360 using Entity Life History analysis.

Remember the feature that an entity must have the potential for more than one occurrence: lists and catalogues are not themselves entities; each entry may be a candidate entity, though.

Identify relationships

Very small systems (those with a dozen or so entities) pose little problem when identifying relationships. Large systems, though, may include several hundred entities, with necessarily complex webs of relationships.

One way of identifying relationships between entities is the use of a *matrix*. The larger the system, the more advisable this is. It enforces a rigorous examination of every pair of entities for a necessary and direct association. Figure 8.22 gives an example of such a matrix, based on a subset of SS plc.

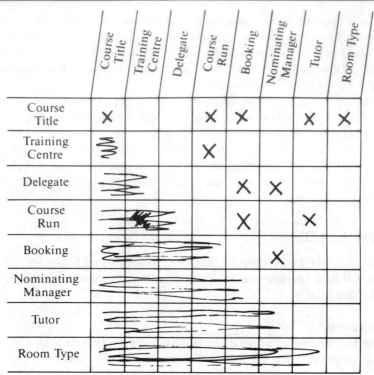

Figure 8.22

The marking of the grid is a matter of local standards. A relationship can be noted by a simple *y* or *x* on the intersection; it can be shown more rigorously by indicating the cardinality of the intersection, with an *m* : *d*, stating which was master and which was detail.

Having identified the relationships, the more difficult tasks of naming each end and agreeing optionality remain. This must be done, as always, with the help of the User.

Draw the Logical Data Structure

The first LDS is drawn in Step 110, but this is a crude overview. Step 140 refines and completes this picture, as the fact-finding exercise progresses. The LDS for the Required System develops during Stage 3, particularly Steps 320, 340 and 360.

There are no hard and fast rules for drawing the structure, except that you must be prepared to make several attempts before you and the Users are satisfied. There are a few guidelines, however, to simplify the process.

- Those entities that have the most relationships should be placed at the centre of the page.
- Those entities that have no masters (reference entities) should be at the top of the page.
- Let natural hierarchies of relationships cascade down the page.
- Try to avoid crossing lines: logically there is nothing wrong with them; cosmetically, they can make the model hard to read.
- Name all relationship ends on the model.

Normalize the LDS

Full normalization (see Chapter 11) does not take place until Step 340. However, to help identify or clarify some entities, it can be done informally in Step 140. The normalization process will not be described here.

Validate the LDM against functional requirements

The LDM must be consistent with the DFD on two counts: processes and data stores.

- *Processes*   Every entity on the LDM must be capable of being created, deleted and, usually, amended. The DFD should show processes that do all of these for each entity.
- *Data stores*   Ensure that after carrying out logicalization of data stores that each entity is represented in only one Logical Data Store.

Once the DFD and LDM have been cross-checked, and one or both amended accordingly, validate the access paths around the model to support the requirements. For this, use the Elementary Process Description, and ensure that the model can provide all the data items to perform each process.

For every EPD on the Data Flow Model there must be an access path on the LDM. This will consist of an entry point into the diagram, such as a key, and a valid navigational path around the relationships. The aim is to ensure that every data item named on the EPD can be accessed—whether to create, update, delete, or read—on the LDS. The data is accessed through one or more of the following:

- Read entity directly (using given key).
- Read next detail from found master.
- Read master from found detail.

If it is not possible to access all the items, either the EPD or the LDS is wrong. Consultation with the User is necessary to clarify this.

This activity is carried out informally in Steps 140 and 150 for the current model, and in Step 320 for the required model. A more formal access path analysis is carried out in Step 360.

Remove redundant relationships

After validating the model against requirements, we must identify redundant access paths (relationships). Look for any 'enclosed' structures on the model. Figure 8.23 is an example of just such a structure, and illustrates the relationships between Course Title, Course Run, Delegate Booking, Delegate, and Booking.

We have three enclosed structures in this subset; all need to be examined. Let us take the requirement to Send Joining Instructions to Training Officers of all Delegates for a forthcoming Course Run.

Access is on Course Run, retrieving all those whose Start Date is the same as the target date. For each of those Course Runs, retrieve each Booking.

The last access can be achieved in three ways: via Delegate/Booking to Training Officer, or Delegate Booking to Delegate to Training Officer, or straight from Booking to Training Officer. Information is needed on Delegate Booking, Delegate and Training Officer to send out the Joining Instructions, so in any event we must find all three entities.

In each access, the same information is available, so is there any advantage in having three access paths? The answer is no, and so we delete one path. In this case

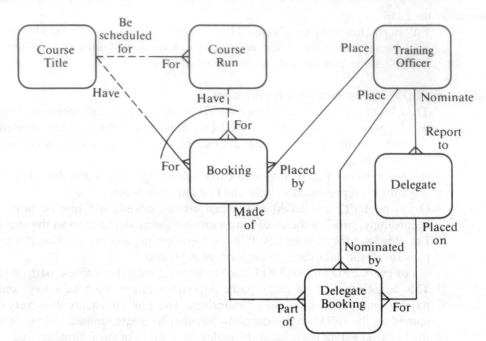

Figure 8.23

the link between Training Officer and Delegate Booking serves no obviously useful purpose, and so we can dispense with it. In removing it we lose no information. The same information is obtained from either of the other paths.

Similarly, we can ask if either of the other two access paths is redundant. We can see that by going from Booking to Delegate Booking to Delegate to Training Officer we can get all the information we need. Going from Booking to Training Officer and down, on the other hand, does not identify which delegates require Joining Instructions for the given course. All we can identify there is that the Training Officers have some Delegates reporting to them, and these Delegates are booked on to some courses.

The relationship between Booking and Training Officer, therefore, is the more dispensable of the two. So it goes.

After this process, our structure is simpler, and looks like Fig. 8.24.

One word of warning: not all enclosed structures contain redundant relationships; all are candidates for such examination, but some need all the relationships to fulfil the requirements.

Beware of what is known as the 'connection trap', or the 'body of the crow'. SS plc does not present us with such a problem, so we shall look instead at a situation in Jobs-For-U (see Fig. 8.25a).

If we are meeting the enquiry: 'What skills does a given vacancy require?', we find that from the structure given, we cannot find the answer. Our point of entry is Vacancy, our next entity is the master Employer, from which we can derive a number of Skill Area, and for each a number of Skills. Employer is the 'body of the crow', because we are travelling up one of its legs and down the other! What this will not do

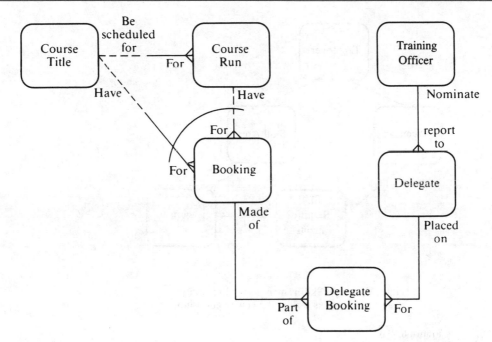

Figure 8.24

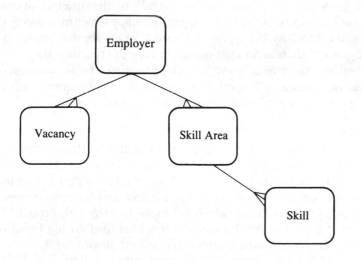

Figure 8.25a

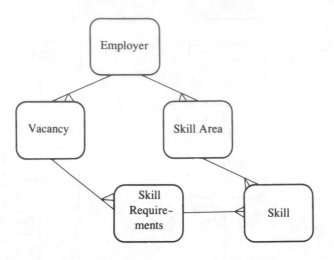

Note: Vacancy—Skill is m:n, so the link entity
is also required, as well as the new relationship.

Figure 8.25b

for us, though, is identify just one Skill that relates specifically to the Vacancy. To meet this requirement we must put in another relationship between the two detail entities, as in Fig. 8.25b. In this way, we travel directly from the entry point to the required information.

You can recognize the 'connection trap' in this case by the structure of one master linking two separate details. If the requirement in such a structure is along the lines of, 'Given a specific detail (a), can we find the specific detail (b) that relates to it?', the answer is always no, if there is no relationship linking the two directly.

Be wary, then, of rationalizing every enclosed structure—in this instance the structure would be necessary. The rule is to examine all the requirements to make sure that you are not losing information by removing a relationship.

Define Enquiry Access Paths

We have already validated the model against the Functional Requirements; in Step 360 we validate the Required LDM against the Enquiry Function Definition. To do this, we prepare a new set of diagrams, called Enquiry Access Paths. This technique of validation is more precise than the earlier validation, and formally defines a set of documented Enquiry Access Paths, which are input to Step 530, Produce Enquiry Processing Models. Each Enquiry Function that is identified during Function Definition (Step 330) has an associated Enquiry Access Path drawn for it.

Enquiry Access Paths have their own diagrammatic notation, loosely based on Jackson techniques, that outline the access to given entities to satisfy requirements. The analysis is carried out on the Required LDM.

To demonstrate the procedure, let us use the following enquiry from SS plc: which courses has a given delegate attended over the last three months, and what were the start dates? The procedure is as follows:

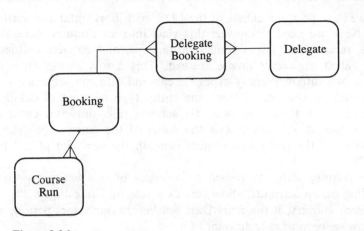

Figure 8.26

1. Draw the part of the LDS needed for the enquiry, either because it contains the information, or because it is needed as a navigation point across the structure to the data needed. Do not include relationship labels on this extract.

    In this instance, the data items needed on the report are: Delegate Name, Course Code, Course Title, and Start Date. Course Code and Start Date can both be retrieved from Course Run entity, while the Course Title is obtained from the Course Title entity. To get to Course Run, we need to retrieve all the Delegate Booking entities, thence to Booking, and then access our required data items from the appropriate Course Run (see Fig. 8.26).

2. For each enquiry that accesses entities across a relationship, redraw the topology to:

    (a) Show each access from master to detail vertically.

    (b) Show each access from detail to master horizontally.

    This is to enable the transformation from LDS to Jackson-style diagram. It does not affect the logical of the model, just eases the analyst into the next step. The results are shown in Fig. 8.27.

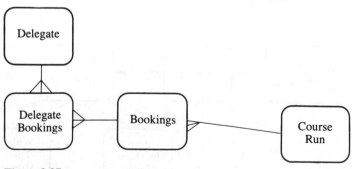

Figure 8.27

3. What we have now is a subset of the LDS, with horizontal and vertical access paths. Next, we need to convert this view into an Enquiry Access Path. To do this, remove the 'crows foot' relationships, and connect entities with an arrow, called an *access arrow*, instead. This arrow shows an access path between two different entity types. It can indicate the sequence of accesses carried out, or also access from one entity type to a set of details; this set will be denoted as an iteration. To achieve this, draw an extra box, and denote it 'set of *X*,', where *X* is the name of the entity type. The iteration will consist of the entity occurrences beneath the new 'Set of ...' box, as in Fig. 8.28.

    If an enquiry wants to regard occurrences of one entity type differently, depending on an attribute, show this as a selection (see Fig. 8.29). Thus, for the present enquiry, if the Start Date fell before our target period, we ignore it, otherwise we must take account of it.

    Mark the entry point for the enquiry with the criteria, e.g., the key, or search criteria (see Fig. 8.30).

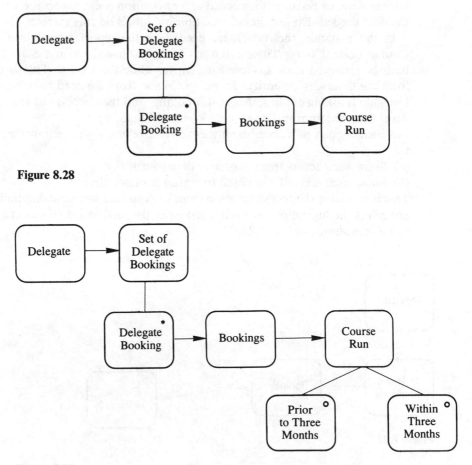

Figure 8.28

Figure 8.29

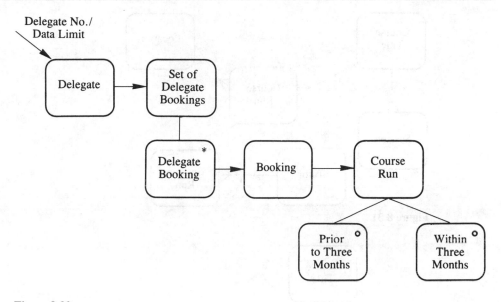

Figure 8.30

4. Confirm that the required data can be obtained in one of the following three
    ways:
    (a) Read record directly using entry attributes.
    (b) Read next detail from current master.
    (c) Read master from current detail.
    If the enquiry requires that occurrences of an entity, or piece of information, are
    presented in a particular order (e.g. date order), and that order is not met by the
    access sequence, make a note on the EAP that a sort process may also be
    required. This will be noted during Create Enquiry Process Models, in Step 530.
Of course, in the example above, it is possible that the entity Delegate Training
History might have given us most of the information; without knowing the attributes
held, it is not possible to say, but for the sake of demonstrating the technique, we
have travelled the long way round!

Let us take another example from SS plc: generate the timetable for a Course Run.
Sessions will be generated from standing data about the course. Tutors for each
session will be named on the timetable; any qualifications they hold (PhD, MBIM,
FBCS, etc.) will be retrieved from their file and entered by their name on the
timetable. Location address and room numbers will also be entered as header
information on the timetable. Figure 8.31 shows us the required view for this
function.

Some of these, such as Session Run can be retrieved more than one way. The access
arrows will follow the actual navigation path rather than reflecting every existing
relationship. Figure 8.32 shows us the master and details relationships converted to
horizontal and vertical access paths.

Now we redraw this section of the model, marking the accesses with arrows and the
enquiry trigger, as in Fig. 8.33. We can confirm that all of the enquiry elements are
met by direct read, master-to-detail access or detail-to-master access. This completes

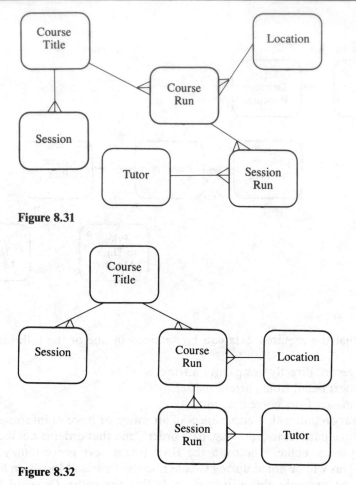

Figure 8.31

Figure 8.32

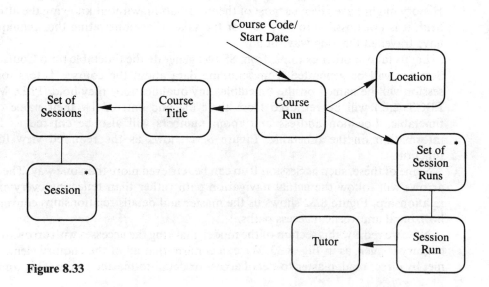

Course Code/
Start Date

Figure 8.33

the EAP, until we reach Step 530, Define Enquiry Processing, when we convert it to a Jackson-style structure, and then into an Enquiry Processing Model.

Presenting the LDM to the User

Quality assurance reviews are not built into SSADM as such, although they are inherent in the Project Management method that drives the SSADM project. However, it is important that the User accepts the LDM as accurate and valid before the formal QA procedure. We will, therefore, show it to the User informally for comment and confirmation. As most SSADM systems will be very large, only show the Users a subset at a time, rather than the whole model: remember, it is a tool for communicating with the Users as well as an aid to our own understanding, so do not give them something too complex and superfluous for the immediate purpose.

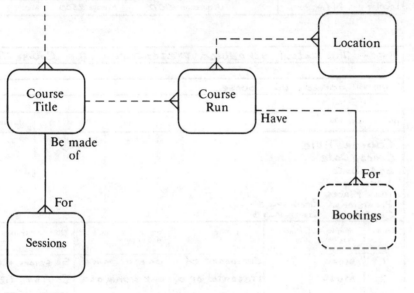

Figure 8.34

Although at each presentation you will show a subset of the model, the entities shown will have other relationships that are irrelevant to the presentation. Show the connected entities as broken boxes, just to show that, although outside the scope of the subset, they are still there in the large model (see Fig. 8.34).

Document the LDM

So far only the Logical Data Structure diagram has been discussed. The full Logical Data Model comprises the LDS, Entity Descriptions, Relationship Descriptions, Attribute Descriptions, and Domain Descriptions. Relationships Descriptions comprise two sheets, one for each end of the relationship.

As the model in detail, these descriptions too must be completed. The descriptions are on standard SSADM forms, and incorporate both functional and operational descriptions.

Entity and Relationship Descriptions, and volumetrics, begin in Stage 1. Full

Entity Description — Part 1

Variant: Current

Project/System SS plc	Author	Date	Version	Status	Page of

Entity name Course Run	Entity ID

Location N/A	Occurrences 1000	Average 2500	Max.

Description The actual scheduled presentation of a Course Title

Synonym(s) Scheduled Course

Attribute name I/D	Primary key	Foreign key
Course Title Course Date End Date Location Max Places Provisional Bookings Confirmed Bookings	✓ ✓	✓

Ref no.	'must be'/ 'may be'	'either'/ 'or'	Link phrase	'one and only one'/ 'one or more'	Object entity name
1	must		Composed of	one or more	Session Run
2	must		Presentation of	one & only one	Course Title
3	may		Allocated to	one & only one	Location
4	may		Subject of	one or more	Booking

Notes

Figure 8.35a

Entity Description — Part 2

Variant: **Current**

Project/System **SS plc**	Author	Date	Version	Status	Page of

Entity name **Course Run**	Entity ID

User role	Access rights
Course Scheduling Bookings	I, D R, M

Owner **Course Manager**

Growth per period

Additional relationships

None - all shown on LDS.

Archive and destruction

Archive 3 months after course ends.
Destroy 18 months after Course Run ends.

Security measures

No special requirements other than access
limitations above.

State indicator values

1. Course scheduled to run.
2. New Provisional Booking made.
3. Provisional Booking confirmed.
4. Provisional Booking cancelled.
5. Confirmed Booking cancelled.

Notes

Figure 8.35b

Relationship Description

Variant: **Current**

Project/System **SS plc**	Author	Date	Version	Status	Page of

Entity name **Course Run**	Entity ID ⟋

Mandatory ☑   Optional ☐   % Optional ☐

Link phase **must be for**

Description **Describes occurrences of presentations of Course Title. Describes Start and End Dates, Location and Tutor.**

Synonym(s) **Scheduled Course**

Object entity name **Course Title**	Object entity ID ⟋

One (1): ☑ Many (m): ☐	Minimum	Average	Maximum
Cardinality description ⟋			
Growth per period ⟋			
Additional properties ⟋			

User Role	Access rights
Course Scheduling	Write
Course Maintenance	Write
Course Bookings	Read
Owner Course Manager	

Notes

Figure 8.36a

Relationship Description

Variant: **Current**

Project/System **SS plc**	Author	Date	Version	Status	Page of

Entity name **Course Title**	Entity ID

Mandatory ☐	Optional ☑	% Optional **10**

Link phase **may be scheduled for**

Description **The Course Title is described in terms of code, intentions of course, duration, cost, effective date**

Synonym(s) **Course, course Type**

Object entity name **Course Run**	Object entity ID

One (1): ☐	Many (m): ☑	Minimum **O**	Average **4**	Maximum **12**

Cardinality description

Growth per period **In total, approx. 500 per month**

Additional properties

User Role	Access rights
Course Bookings	Read
Course Manager	Write
Course Scheduling	Write
Course Maintenance	Write

Owner **Course Manager**

Notes

Figure 8.36b

Attribute/Data Item Description

Project/System		Author	Date	Version	Status	Page of

Attribute/data item name	Delegate Duty	Attribute/data item ID Delegate ID

Cross-reference name/ID	Cross-reference type
	Entity Description I/O Structure Dialogue

Synonym(s) Del. Duty   Del. ID

Description

The unique identifier for each Delegate who attends any course.
The format is AAA 9999 – Branch Code, Serial Identifier

Validation/derivation

Alpha component must have a match with a Branch Code
in current use.

Mandatory ☑ Default value	Optional ☐ Value for null
Logical format Alpha (3) Numeric (4)	Unit measure N/A
Logical length 7	Length description

User Role	Access rights
Bookings A/C	I R

Owner	Accounts Manager

Standard messages

Notes

Figure 8.37

descriptions are completed in Step 320, when the required LDS is drawn. In Step 360, the final and complete LDM, with all volumetrics and supporting operational documentation, is presented, as well as all functional documentation.

Figures 8.35–8.37 give an Entity Description for Course Run, a Relationship Description for the relationship between Course Run and Booking, and supporting Attribute/Data item Descriptions.

Summary

The Logical Data Model provides us with a second view of the system: the underlying structure of the data, and relationships between logical groups of data.

It is developed first of all, in parallel with the DFD, in Feasibility, then in Stage 1. On selection of the BSO, the LDM is expanded to support new functional requirements, and is further enhanced by Relational Data Analysis.

The Logical Data Structure itself provides us with a simple diagrammatic notation, but it must be rigorously supported with documentation describing fully entities, relationships, attributes, and domains.

One point made earlier in the chapter cannot be overemphasized: the notation of the technique often leads inexperienced data analysts to believe that Logical Data Modelling is simply a matter of drawing boxes and lines. It is not. Data analysis is about understanding the nature of the data in the environment, about correctly classifying it according to the sphere of interest, understanding its use, and recognizing and resolving problems associated with the representation, such as recording many-over-time.

EXERCISES

1. The following text describes a college library environment.

 A college library holds books for its members to borrow. Each book may be attributed to one or more authors. Any one author, of course, may have written several books. Up to 10 copies may be held of popular titles.

 A member may borrow up to six books at a time. If books are not returned on time a fine will be levied. If the books are not returned by a week after a third reminder, and fines are not paid, the member may be put onto a blacklist until the position is rectified.

 If no copies of a wanted book are currently in stock, a member may make a reservation for the title until it is available.

 (a) Draw a Logical Data Structure for this environment.

 (b) Draw Enquiry Access Paths to fulfil the following requirements:

 (i) List all members who have reserved a particular book.

 (ii) List all members on the blacklist.

 (iii) List all the books written or co-written by a specified author.

2. The text below describes the environment that exists within a rail transport authority.

 The authority divides its services into lines (e.g., cosmopolitan line, peripheral line). Each line is served by a number of depots (a depot serves only one line) at which trains, drivers, and guards are based.

 Each line has many stations, and because each line crosses all other lines at some point in the network, a station may be on many lines.

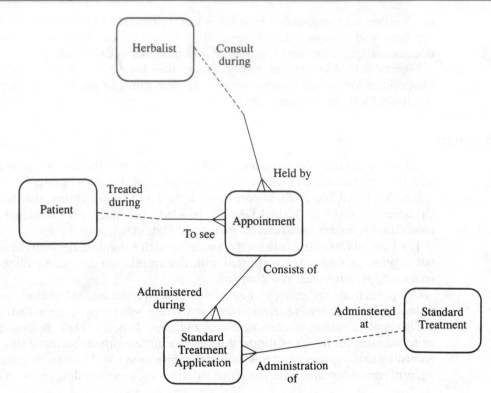

Figure 8.38

Lines may have branches at either end, so each line may have many routes. A route is described as running between two stations, in one direction only. Thus, route 1 might be from Station A to Station M, while Route 7 might be from Station M to Station A. A route will not embrace more than one line.

Drivers and guards work on trains based at their depot only. Every journey is assigned a driver, but on some journeys, guards are not considered necessary. A journey is a particular occasion of a train travelling a route.

(a) Draw a Logical Data Structure for this environment.

(b) Draw the Enquiry Access Path to meet the following requirement: identify all trains, drivers, and guards assigned to a specified line.

3. LOITER-WITHIN-TENT is a holiday company that specialises in providing caravan and camping sites in popular tourist resorts. Each site has a number of plots that can be booked by the week or by the fortnight. Each plot is identified by a number, and may be equipped for either a caravan or a tent. Plots with caravans have extra facilities, such as electrical points and water points. Tent plots do not have these facilities.

Holiday-makers may make a Booking for a plot, each Booking being for a multiple of days. The Invoice will be issued against a Booking, regardless of numbers in the party, although details will be kept of the individual responsible for the Booking.

Invoices may be paid off in one go or, if more appropriate, may be paid in a fixed number of instalments.

Draw a Logical Data Structure for LOITER-WITHIN-TENT's Booking and Invoice system, based on the above information.

4. Figure 8.38 depicts an LDS for a Herbalist clinic. Each time a Patient is treated at an Appointment, a record is made of the Appointment, and the treatment that has been administered. For any one ailment or condition, there are a number of alternative Standard Treatments available. Following a number of patient complaints, there is a requirement for a Report to be produced which lists all the treatments administered for a specific ailment during the previous week. Draw an EAP to meet this requirement.

9. Data Flow Modelling

9.1 Aims of chapter

In this chapter you will learn:
- Where DFDs are used in SSADM
- How to construct a Current System DFD
- How the Level 1 DFD can be decomposed to Levels 2 and 3 DFDs
- How the Current Physical Model is logicalized
- What documentation is needed to support a DFD

9.2 Where Data Flow Modelling is used in SSADM

SSADM creates and amends DFDs in the following steps of analysis.

Step 110—Establish analysis framework This step uses the results of any feasibility report to verify the Project Initiation Document. Any discrepancies between these and the current situation are amended by updating the DFDs.

Step 130—Investigate current processing Here we use the DFD as an aid to fact-finding and recording the results of the investigation. The overview DFD produced earlier is reviewed and, if necessary, extended in the light of the investigation. As the level of detail grows, so the DFD must be 'decomposed' to its lower levels—Level 2, or maybe even Level 3.

Step 150—Derive logical view of current services The physical DFDs created above must be converted to give a *logical* view of the system, see Sec. 9.7. If there is no current environment the analysis begins with this diagram.

Step 210—Define Business System Options To help the User select a business solution from a menu of possible solutions, the current view must be adapted to give a required view, see Sec. 9.8.

Step 220—Select Business System Option Once the User has selected the option, Level 1 and Level 2 of the critical processes may be drawn to support it.

Step 310—Define Required System Processing Using the Logical DFM, construct a full Required DFM from the selected BSO and the Requirements Catalogue.

Step 330 ⎫ *Derive system scope* If any extra updates or data flows are identified
Step 360 ⎭ during these steps, the DFM is amended to reflect these.

9.3   Data Flow Diagrams

Data Flow Diagrams are the core of most methods described as 'structured'. They come from the work of the Yourdon Corporation, and early structured analysis and design methods revolve around them. Yourdon, De Marco, and Gane and Sarson's methods all use DFDs, each with its own notation and characteristics.

The virtues of DFDs are a result of their simplicity: the notation consists of just four elements, described below.

After we have carried out our initial investigation of the current system, we need a way of describing its activities that we can show to the User to confirm or correct our understanding of the system. Traditionally, the two possible methods were prose narratives or technical descriptions. Both of these were unsuitable.

Narrative is prone to ambiguity. With the greatest care in the world, the analyst may still describe a work area in terms that the User seems to understand, but which are, in fact, inaccurate. Any mistakes that emerge in reviews may take considerable effort to correct, and risk introducing further errors and ambiguities.

What is perhaps a greater drawback to narrative, however, is that it is necessarily long-winded and linear in structure. Thus it is harder to read and assimilate, and does not present a fair view of a system in which many things happen concurrently.

Technical descriptions are meaningful to those people who are involved in a given functional area, but less so to others. As the analyst is likely to investigate several functional areas in one study, there is a danger that several standards will dictate the contents of one document, and several different jargons will be incorporated. This is not acceptable; it makes comprehension of the overall document harder, and amendment more fraught.

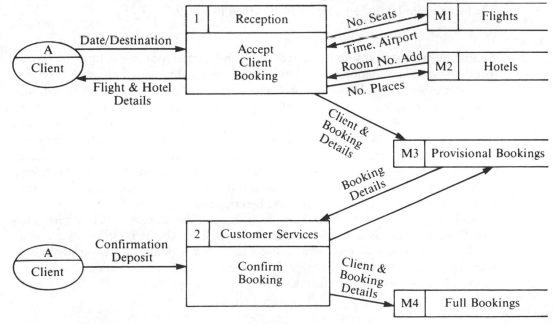

Figure 9.1

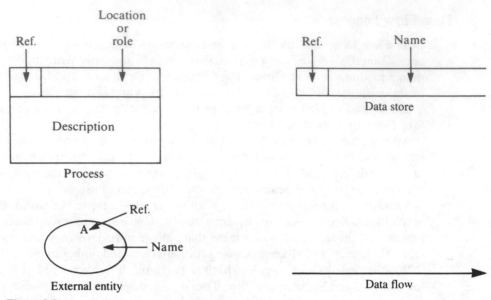

Figure 9.2

What is required is a simple, comprehensive notation that gives an overall view of the organization as well as appropriately detailed descriptions of activities. It should be easy and quick to draw, and as easy to amend.

Figure 9.1 shows a simple DFD describing the Current Physical system in a Travel Agency, receiving a request for a Booking for a given Flight and Hotel. There are just four elements in this diagram: processes, data stores, data flows and external entities. The meaning and structure of each is described in Fig. 9.2.

Process

A process is an activity that receives data and carries out some form of transformation or manipulation before outputting it again. The activity may be carrying out calculations, creating a new document from information that triggered the process, or amending the document that entered our system. It is depicted by a box divided into three parts: the upper left position is given a number. This has no significance at all other than as a reference number; it does not imply priority or sequence. As a reference, for user communication, however, it is an important feature. The longer rectangle beside it names the location where the processing takes place; this may, on an overview DFD, be a broad term, Sales Accounts, etc. As the DFDs become more detailed so do these descriptions.

The rest of the box describes what is happening in the process. The rule here is to keep the description as terse and meaningful as possible. Use an imperative verb with an object, but make the verb specific. 'Process...' and 'Update...' are too vague and give little clue as to what is meant. 'Calculate...', 'Add...' or 'Validate...' give a clearer picture of what is happening.

Data store

A data store is a place where data comes to rest. It may be a filing cabinet, an in-tray, a card-index, a reference book, or a computer file. Anywhere that data is stored and retrieved is called a data store.

The notation is simple: a long, open-ended rectangle, with a box at the left-hand end. The box is labelled with an alpha prefix and a number. The alpha is either D (for an automated data store) or M (for a manual/card data store). As with the processes, the number has no significance; it is purely a reference. The rectangle is labelled with a description of the contents of the data store.

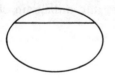

Figure 9.3

If, for the sake of tidiness in the diagram, you wish to show the data store in more than one part of the diagram, draw a bar beside the left-hand box, as in Fig. 9.3. Each occurrence of the data store concerned will display that box.

External entity

The third notation, a lozenge, represents an external entity. External entities are those bodies outside the system boundary which interact with the system. They may be external to the whole company, such as customers, Inland Revenue, Customs and Excise, or just external to the application area. Thus if we are modelling a Sales Office system, Accounts and Despatch areas would be shown as external entities. Each external entity communicates in some way with the system, so there is always a flow of data shown between a process in the system, and an external entity.

The entities are labelled with a singular noun describing the role of the entity, e.g., Accounts, VAT Office, Credit Manager. Above the label is an alphabetic character, again for reference purposes only.

Figure 9.4

As with data stores, it may be desirable for the sake of clarity to duplicate an external entity on the diagram, rather than have arrows from all points converging on one entity. If that is the case, put a small line along the top of the lozenge, as in Fig. 9.4.

Data flow

A data flow represents any passage of data into the system, out of the system, or between elements inside the system. It is represented by an arrow between the source and recipient of that data flow. In the world, it may take the form of a standard document with fixed content, or a telephone call. It may be an enquiry, a functional document, or a memo. Wherever traceable data is passed, it must be shown by an arrow. At the highest level DFD, one arrow may represent several data flows, which may be decomposed into the individual flows at the lower levels. Thus in our example, SS plc, a single data flow at the top level from Client to Bookings may read 'Booking'. As we expand and decompose the diagram, that one flow may decompose into Enquiry, Provisional Booking, Confirm Booking, and Cancel Booking.

There are some validation rules about where data flows may or may not travel:

- Data stores may not be linked by data flows: flows must travel from one to another via a process.
- External entities may not send or receive data flows directly to or from a data store: they must communicate via a process.
- Data cannot be generated by a process, or be swallowed by a process; documents may be swallowed or generated, but there must be output that is related directly to all inputs to the process.

The technique is called Data Flow Modelling, suggesting that there is more than a simple diagram involved. The term 'Data Flow Model' refers to the set of DFDs with the supporting documentation, described later in this chapter.

Use in SSADM

Inputs to DFM are as follows:

- Project Initiation Document
- Feasibility Report
- Current Physical DFM (for Logical DFM)
- Logical DFM (for Required System DFM)
- Requirements Catalogue (for Required System DFM)
- Selected Business System Option (for Required System DFM)

Outputs from DFM are as follows:

- Data Catalogue
- Data Flow Model
- Logical Data Store/Entity Cross-reference

Section 9.4 looks at how to construct a DFD for our case study, and how to convert this physical description of the current system into a diagram of the new, required system.

9.4   How to begin

The high-level DFD produced at Step 110 identifies the system boundary, the external entities, data flows across the boundary, and the principal functional area/activities within the boundary. If the system is a particularly complex one, or the DFD technique is new to you so that you are not sure how to start, the following will give you a guide to producing that first high-level diagram.

The first thing is to establish a *context*, and to that end we can produce a *Context Diagram*. We then transform that Context Diagram into a Document Flow Diagram, and from there produce our Level 1 DFD.

Context Diagrams

If the high-level DFD is known as a Level 1 DFD, the Context Diagram can be regarded as a Level 0 DFD. This can be drawn by following the steps below. To illustrate the techniques, I shall use the SS plc case study.

Step 1   List the documents used in the system.

SS documents:

1. Bookings Form
2. Joining Instructions
3. Confirmation of Booking
4. Delegate List
5. Newsletter
6. Course Cancellation Notice
7. Transfer Fee Request
8. Payment/Advice
9. Monthly Stats. Reports
10. Delegate Cancellation
11. Tutor Allocation
12. Tutor Availability

Step 2   List all the sources and recipients of these documents.

SS sources/recipients:

1. Bookings (BK)
2. Billing (BI)
3. Course Maintenance (CM)
4. Course Scheduling (CS)
5. Course Managers
6. Tutors
7. Delegates
8. Branch Training Officers
9. Accounts Department

Step 3   Draw a box representing the system and show the flow of documents from these sources and recipients, as in Fig. 9.5. Those areas which are known to be inside the system (here they are CM, CS, BK, and BI) are hidden inside the box.

Document flows

From this Context Diagram, or Level 0 DFD, we can expand the diagram into a Document Flow Diagram by leaving out the 'black box', and showing the documents' path between the entities, as in Fig. 9.6.

9.5 Developing the Physical Model

Top-level DFDs

Having drawn this document flow, we must, with the help of the Users, agree on the boundary. This can be marked by a dotted line, for ease of reference. Those entities outside the boundary are now known to be external entities. Those inside the boundary are functional areas that act on the flows of data. The first thing to do at this stage, then, is to transform those entities within the boundary into processes, and label them accordingly, as in Fig. 9.7.

Our high-level DFD is not complete yet: we have not identified how many processes each of these functional areas performs, and what data stores are used. As

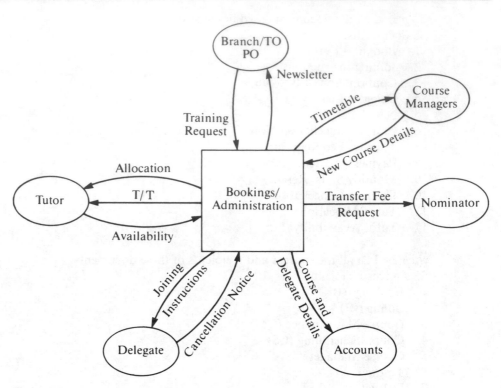

Figure 9.5

we are describing a physical system, we must consider the types of data moving through the system. As a rule-of-thumb guide, we can consider the following types of data store:

1. Standing data, used for the day-to-day functioning of the system and kept up to date. Such items would be Course Brochures, Course Schedules, and Hotel Lists.
2. Historical data that is required to be maintained for reference and enquiry purposes, but is no longer 'live'. Such items could be details of Course Presentations that are completed.
3. Temporary data stores, such as collections of delegates who require Joining Instructions to be sent out the next day, their details being batched together in the previous afternoon in readiness. Once the Joining Instructions are despatched, this collection of data will not be needed and the data store will no longer exist.
4. Extracted data that is retrieved from different sources for the purpose of preparing reports, statistics, and so on.

What we must do at this point is to examine the data flows across the boundaries into the processes/functional areas on our 'first cut' attempt, and see what actions are performed on them. Each process will probably require at least one data store. Our initial investigations will identify all such data stores and their access, so whatever is unclear at this point can be clarified then.

The labelling of the processes will help us to sharpen our understanding of what is

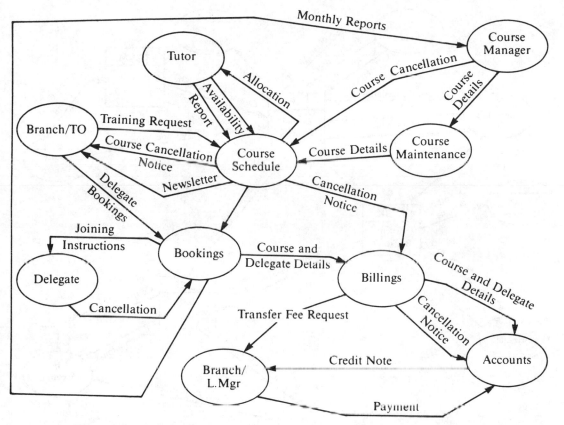

Figure 9.6

happening in each area: the process should be identified by a clear, unambiguous verb with a direct object.

If all we can produce is 'Update Wait. List' we must break that down, now or later, into some such activity as 'Place New Applicant on List', 'Remove Applicant from List', and so on. In this way, we expand our initial attempt to give a clear, high-level description of what happens to our data flows in each section, and what data stores require to be accessed. Figure 9.8 gives us a view of the system after this process.

While this seems to be an accurate view of the organization's operations, it is not easy to read. As one of the virtues of the DFD is its use as a communications tool, we must try to tidy it up and make it intelligible.

There is no hard-and-fast rule for this, just several redraws. A drawing tool, or even better, a CASE tool, would best serve us here. In terms of saving time, automation gives benefits, and in terms of preserving the integrity of the diagram during redraws, CASE can ensure that data flows stick to each object as it is moved about the screen, so that the logic of the redraw is the same as the logic of the original, messy draft.

Two possible 'fixes' to help tidy the Level 1 diagram are combining related data flows, and related external entities. For example, we have two external entities

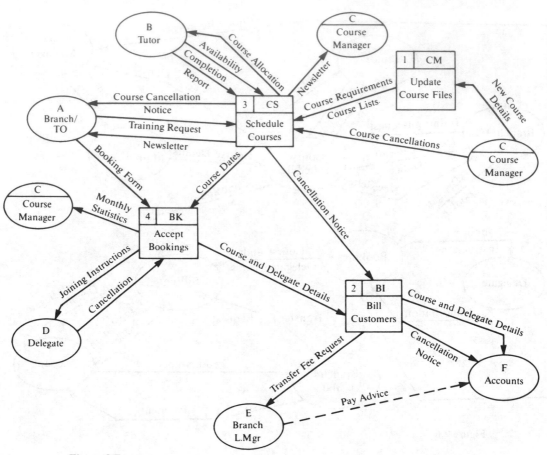

Figure 9.7

relating to the Branches: Branch/TO and Branch L.Mgr. At Level 1 it is permissible to combine them to form one Branch, as in Fig. 9.9.

On our DFD, we have three data flows coming from external entity Branch/TO into Process 4, Receive Application. At Level 1, these can be combined to form one composite data flow: Booking, as in Fig. 9.10.

Optimizations such as these will tidy up the Level 1 diagram, but must be decomposed again at lower Level 2, to ensure that we have modelled the activities in detail, and have not overlooked some important flow, or User Role.

Having carried out our tidying up operations, Level 1 of SS plc now looks like Fig. 9.11.

Validation of DFD

Before we move on from this initial high-level DFD, drawn up at Step 110, we must make sure that it is consistent. Below is a checklist of points to watch before moving on to the detailed investigation which will take us to the lower levels.

1. Has each process a strong imperative verb and an object?

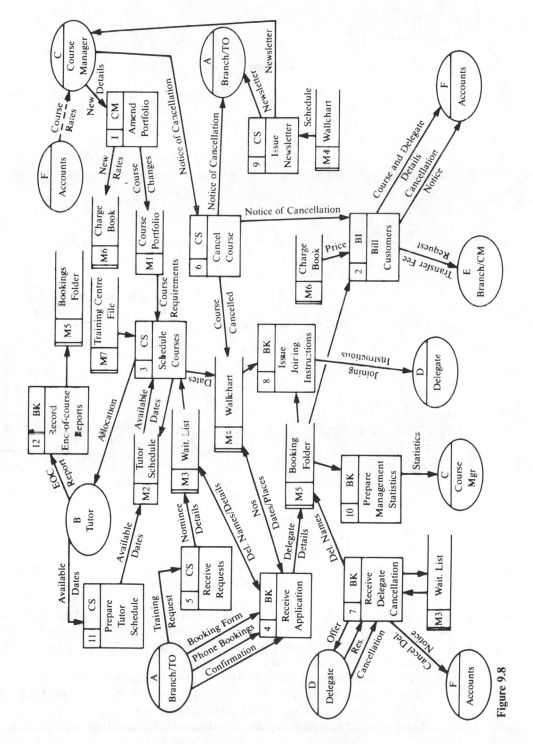

Figure 9.8

127

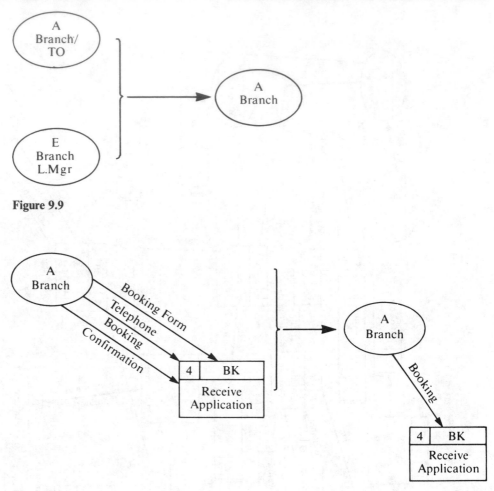

Figure 9.9

Figure 9.10

2. Are data flows in related to data flows out? Data should not be swallowed up by a process, only transformed in some way. A data store is the only place data is allowed to rest. Similarly, data cannot be generated by a process. A document may be, but the data on the document comes from a data flow into that process.
3. Can the flows be reduced? If a process is too busy, it can perhaps be broken down into two or more processes: six data flows in or out of a process should be sufficient.
4. Do all data stores have flows both in and out? A one-way data store is of little use, unless it is a temporary data store, or it is a reference file only. If the Current Physical DFD should identify such a data store, confirm with the User that you have correctly understood the procedures.
5. Are symbols correctly labelled and uniquely referenced?
6. Do all external entities communicate with a process? No entity should be allowed direct access to data, either to read or to update it.

When you are satisfied with these questions of DFD notation and logic, it is time to

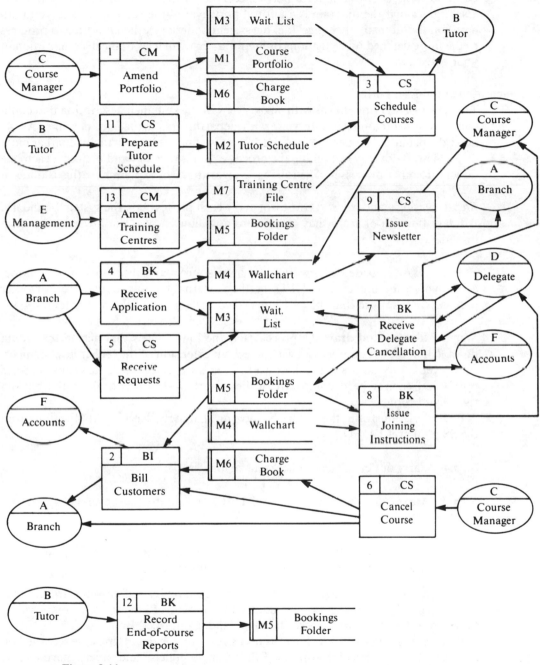

Figure 9.11

check with the User that it accurately reflects the business system logic as well. If the User agrees that the diagram is a faithful portrayal of the work area, and you both agree on the 'domain of change' (i.e. the system boundary), then it is time to progress to the more detailed investigations of Step 130, that will lead to the decomposition of your overview.

Decomposition of top-level DFDs

The Level 1 DFD presents us with an overview of the system, a description that could come from a preliminary interview with departmental managers, perhaps. As we delve deeper into the operations of our system we find we have to include rather more detail. This is done by examining the processes in more detail and breaking each one down into other processes. The following algorithm will explain how this is done.

Step 1   Using standard SSADM form make each process box the system boundary. All data flows to or from that process are now flows across the lower level system boundary.

Step 2   Draw, outside the new boundary, the sources and recipients of these flows, as shown on the higher level DFD, be they external entities, data stores, or other processes. Make sure that they are labelled consistently with the higher level.

Step 3   Identify and draw the processes at the lower levels that act on these data flows. Number the subprocesses with a decimal extension of the higher level number, i.e., Level 1, Process 3 will break down to Processes 3.1, 3.2, 3.3, etc. Those process boxes that cannot be decomposed further, mark with an asterisk in the bottom right-hand corner.

Figure 9.12 illustrates these steps by looking at activity 'Receive Application' from the SS plc Level 1 DFD.

Step 4   Carry out consistency checks, as before.

Step 5   Make sure that all lower level DFDs map onto the Level 1 diagram, by checking dataflows.

Step 6   Review the lower levels with the User to be sure that you have depicted every activity actually performed in the system under investigation.

You may find that a single data flow on the top level is itself decomposed at Level 2. Thus, in Fig. 9.12, the Level 1 data flow 'Booking Request' is decomposed in Processes 4.1 and 4.5 into 'Fresh Booking', and 'Confirmed Booking'.

When you have taken your DFD as far as you can, the details must be recorded on an Elementary Process Description (EPD) using a concise and precise narrative. If more than four or five sentences are required, perhaps the process has still to be broken down to another level.

If a process involves making a decision, this is to be recorded on the EPD, not on the diagram. (Note: a common mistake made by people learning the DFD technique is to treat the chart as a flowchart: this is *wrong*! Use a tool such as a decision table, or

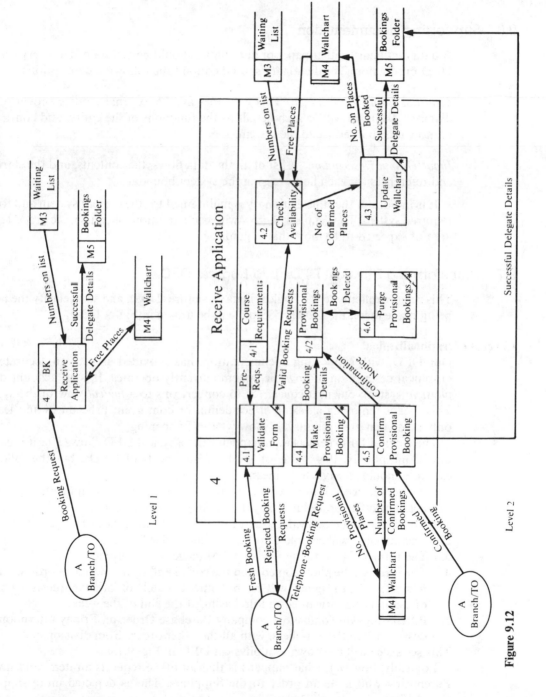

Figure 9.12

131

decision tree to describe decisions.) If the decisions are complex, i.e. there are more than five or six possible courses of action, the process will probably be too busy, and so capable of further decomposition.

9.6   Supporting documentation

A data dictionary, either paper or automatic, should be maintained at every stage of DFD production. The dictionary should contain the following descriptions.

External Entity Description   This describes all the external entities shown on the diagram. Included will be such details as the functions of the entity, and constraints on how it is to interface to the system.

Input/Output Description   A list of all the data flows, the contents, and the start and end references for each flow crossing the system boundary.

It is important that this dictionary is maintained for the current system and for the required system. The details will grow with each iteration, of course; the first attempts are not expected to be more than a guide.

9.7   Transforming Physical DFDs into Logical DFDs

This refers explicitly to the Logical Data Structure (LDS), and so to clarify the points being made there, the LDS for SS plc will be drawn in Fig. 9.14.

Concept of 'rationalization'

Our DFD, with supporting documentation, has provided us with an accurate and graphical description of how the system currently operates. It is an account of the *physical* system; our next concern is to convert this to a *logical* view of the system. This means removing all signs of accidental or contingent procedures and leaving only those that are necessary to show what is happening.

The Logical DFD depicts *what* happens; the Physical DFD shows *how* it is carried out. Conceptually, this is one of the hardest aspects of DFDs, but the following example should clarify what I mean.

Assume that a clerk in an office requires a new piece of equipment, say a filing cabinet. The office procedures may dictate the following pattern:
1.   The Clerk concerned will request the item from the Office Manager.
2.   The manager will complete an Equipment Order Form.
3.   The form will pass to the Divisional Manager for approval.
4.   The form will be photocopied, and the copy kept for divisional records and audit purposes. The original will be put into a batch of other requests from the division, to be forwarded to Purchasing at the end of the week.
5.   Purchasing will complete a company Purchase Order on Friday afternoon and despatch it to the Suppliers, with all the other orders from divisions.

This sequence will be shown as a physical DFD in Fig. 9.13a.

Logically, however, what happens is that an office requests an item, and notifies Purchasing who issue an order to the Suppliers. This is depicted more simply as Fig. 9.13b.

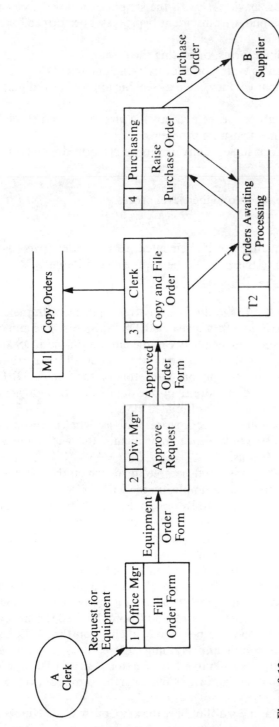

Figure 9.13a

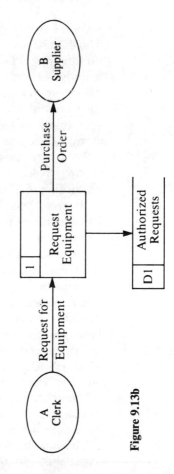

Figure 9.13b

133

This example shows that the object of this exercise is to remove the constraints of the physical work environment, leaving us with the simple data flows necessary to show what is happening. To this end, we must identify processes performed for any of the following reasons:

- *Procedural*   Such as batching forms, or sorting them.
- *Machine/tool-related*   Such as typing or photocopying documents.
- *Political*   To do with company policy, such as seeking approval from particular managers for a procedure.
- *Geographical*   An organization spread over a wide area may need special pro-cedures and data flows to meet ensuing problems.

To carry out this 'idealization', or logical conversion, we perform three tasks:

1. Rationalize data stores.
2. Rationalize processes.
3. Group bottom-level processes into higher-level processes.

Rationalizing data stores

Data stores, in whatever form they are implemented, can be categorized as *main stores* or *transient stores*. Each of these should be approached differently.

MAIN DATA STORES

The data stores on our Logical DFD should have a correspondence with the entities on the Logical Data Structure. Figure 9.14 gives the LDS for SS plc. The aim of this activity is to find a logical correspondence between the entities on the LDS and the data stores on the DFD. The end of the exercise will be to match all the attributes on the LDS with the data items in one and only one data store on the Logical DFD, i.e., such that no attribute will be found in more than one data store. Each logical data store will refer to one entity, or to a group of related entities.

To achieve this, we must examine the data stores and see what purpose each is serving logically. Administrative convenience may dictate the way data is held physically; that no longer concerns us.

By cross-referring to the LDS, we should be able to see which data stores serve which entities or groups of entities. For example, on the LDS for SS plc we find that Course Run is related to Training Centre, to Rooms, to Session Run, to Bookings, to Tutors and to Course Title. Our task is to see which of those entities will be placed in the same data store on the DFD as Course Run details. One way to do this is to examine the way in which entries on the data stores are keyed.

If two data stores on the diagram share a common key, consider whether they are, in fact, the same data store. Here, the data store Wallchart is the only one that is keyed on Course Title/Date Commencing. What is held on the Wallchart, though, is information about the location (Training Centre) and Rooms booked for that course. That will help us make a correspondence between Wallchart on the DFD and the entities Course Run, Room Allocation, and Training Centre.

The other possible entities, Booking, Course Title, Session Run and Tutor, must be similarly examined to see if they belong to the Wallchart information, or to other data store subjects.

The data stores Charge Book and Portfolio share the key Course Code Number. It is a matter of administrative convenience that they are kept separately, so in our logical

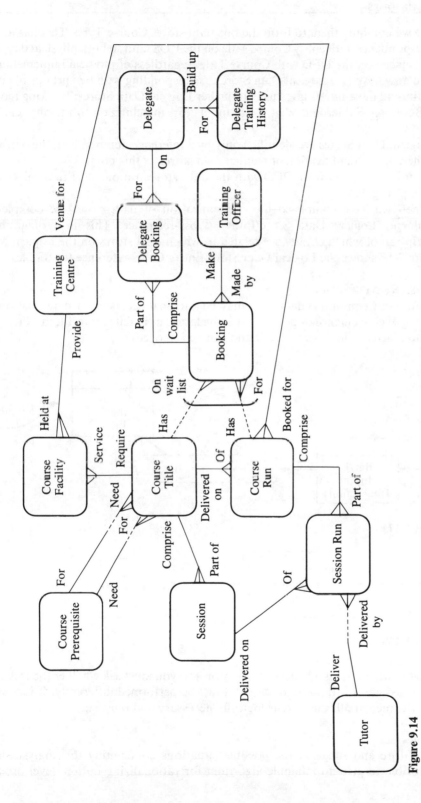

Figure 9.14

135

DFD we combine them to form the one data store, Course Type. This has an exact correspondence with entity Course Title on the LDS, and so both physical data stores are replaced on the DFD with Course Title, regardless of physical implementation.

We now have two Logical Data Stores, corresponding with two groups of entities. Continuing the analysis gives us two further Logical Data Stores: Booking (combining Bookings Folder and Wait. List) and Tutor, matching on to the physical Tutor Schedule.

Logical Data Stores are identified only by a reference number, with the prefix D or T. There is no identification of manual data stores at this point.

Redrawing our Level 1 DFD with the data stores rationalized thus, gives us the model in Fig. 9.15.

When we have completed the rationalization of data stores we complete the document: Logical Data Store/Entity Cross-reference. This is a diagrammatic description of which entities are related to which data stores on the Logical Model. Figure 9.16 shows the Logical Data Store/Entity Cross-reference for SS plc.

TRANSIENT DATA STORE
A transient (temporary) data store may be a logical necessity, or may exist only for reasons of convenience or policy. If it is the latter, as in data store T1, in Fig. 9.17a, it has no place on the Logical DFD and can be removed.

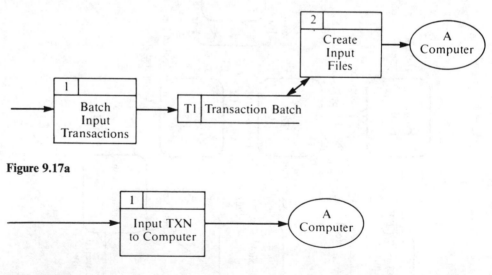

Figure 9.17a

Figure 9.17b

To tell whether it is a logical necessity or not, you must ask whether the task can be performed at all without it, or would it just be performed differently. If it could not be performed at all, the store is logically necessary and must stay.

Rationalizing processes

The nature and range of the possible situations confronting the analyst make it impossible to give an infallible algorithm for rationalizing bottom-level processes;

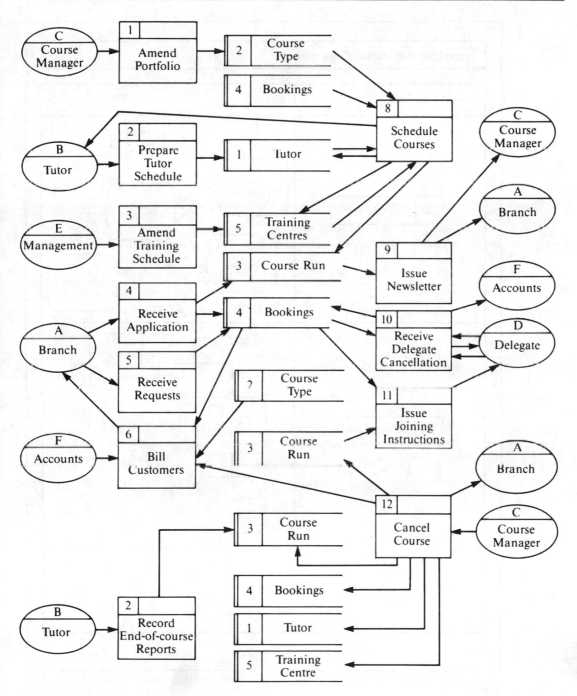

Figure 9.15

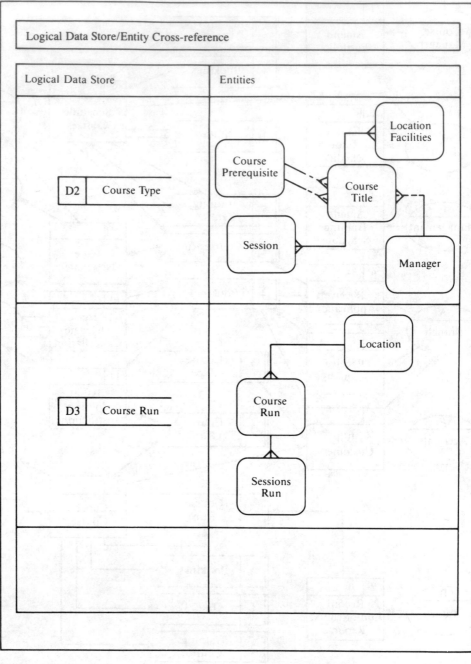

Logical Data Store/Entity Cross-reference

Logical Data Store	Entities

Figure 9.16

judgement based on experience is really the only guide. That is no help, of course, to analysts on their first SSADM project, so below are some guidelines and examples of processes that can and should be rationalized, i.e., merged or removed altogether:

1. Any process that retrieves data purely to print or display need not be shown on the diagram; the Requirements Catalogue should keep a record of it.
2. Any process that merely sorts data, rather than updating it, need not be shown on the diagram.
3. If a process involves a decision being taken by a person, rather than automatically, that process should be split into two or more processes; the person involved in the decision will be depicted as an external entity. For example, in a warehousing system Fig. 9.18a becomes Fig. 9.18b.
4. If two processes are duplicated, examine the whole picture to determine whether or not they need to be represented separately. It may be that you can combine them, but only the complete view will tell you.

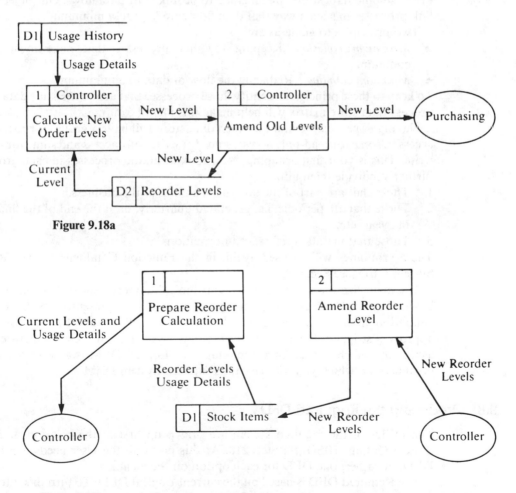

Figure 9.18a

Figure 9.18b

5. If in your decomposition you have produced a sequence of processes, moving from one to the next, to complete one task, these can be combined to just one or two processes that define the overall task.

    As we have removed all reference to the physical environment, so we must amend the labelling to reflect this. Data flows should be labelled with a succinct description of the flow, not with a document name. Remove references to locations on the process boxes: these reflect the physical, not the logical system.

These examples are guidelines to get you started. You may well find that after you have rationalized data stores and processes your diagrams have about three or four fewer boxes than before. As long as all of the symbols reflect the logic of the system and not the physical, or contingent, constraints, that is all right.

Every stage of DFD production is iterative, so this part will also probably require several passes before it is completed to your satisfaction.

Grouping lower level processes

Processes and data stores are all going to be linked by data flows. Our objective is to link processes in such a way that data flows are kept to a minimum.

Two principles to guide us are:
- *Maximum cohesion*   Keeping together processes that have most data in common.
- *Minimum coupling*   Reducing the flow of data to a minimum.

To keep to these principles, identify those processes that access the same data store. A process/data store matrix will help make this identification.

Having done this, we find that each data store will have a group of processes that access it, some for read-only access, some for write-only access and some for read and write. This is your first grouping. Now look at all the processes in each group, and further subdivide them into:

1. Those that are part of the everyday running of the business.
2. Those that are periodic, i.e. generated quarterly, or at the end of the financial or tax year, etc.
3. Those that maintain reference information.

These groupings will be used again, in the Function Catalogue for the Required System DFD (see Sec. 9.8).

The logicalization process is not algorithmic, but more heuristic, or rule-of-thumb based. It is not expected that this product would be presented to the Users for validation: after all, they have already approved the Physical DFM. To ask them to approve a stripped-down version of the same system as an accurate picture may be asking too much of their understanding. The Logical DFM is a necessary working document on which you will base the Required System DFM.

9.8   Producing the Required DFD

The DFDs of the required system are drawn up first in the creation of Business System Options (BSO), in Step 210. At this point all the User needs is a top-level DFD, or rather, one DFD for each option on the menu.

The Required DFD is based on the current Logical DFD. To turn this into a BSO, consider, with the aid of the Requirements Definition, what processes can be automated, which may/must be manual, what data stores will be required, and any new or altered data flows.

It may be that there will not be drastic differences between the options, but they must be sufficient to show the User clearly what the choices are. As shown in Chapter 16, the BSO will comprise top-level DFDs, to show the scope of the new system, rough Cost/Benefit Analyses and an Impact Analysis for each.

Once the User has made a choice—and this selection will probably be a combination of features from two or more BSOs—then the top-level DFD is expanded as with the Current Physical in Step 130. The Logical DFD will be used for reference where possible, to ensure that facilities currently offered are not omitted from the new system, unless through deliberate choice.

When the decomposition is complete, a check should be carried out against the Requirements Catalogue to ensure that all the items on the list have been met by the new processes.

Although the DFM for the current system must be rigorous and accurate, the emphasis should be on development of the required model: SSADM is more concerned with meeting the requirements for the future, than in examining the current ones.

When the Required DFD has been taken to its lowest level, the accompanying documentation must be completed:

- *Data Store/Entity Cross-reference*   See Fig. 9.16.
- *Elementary Process Description*   A brief description of each process, to amplify the labelling (Fig. 9.19).
- *I/O Descriptions*   Every screen, form or print used in the system is to be entered, with data items, format, and size. Details of enquiry screens and reports, as well as updates, will be entered (Fig. 9.20).
- *External Entity Descriptions*   A brief description of each external entity (Fig. 9.21).

When this is completed, review, as always, with the User, to ensure that nothing is omitted, and that nothing is unclear.

Summary

Using DFDs and supporting documentation, we have now analysed the workings of the current system, reduced it to its *logical* functions and described a new system to meet the stated problems and requirements. This sequence can be represented by Fig. 9.22.

All we have done, though, using this very useful graphic tool, is to *describe*, not specify, the system. We shall use the Required DFDs and documentation later in the design process, when we prepare the enquiry processes and update accesses, and when we carry out Relational Data Analysis. It should be remembered that SSADM uses DFDs primarily as a tool for analysis and communication, not for design.

The design process makes use of the view of the required system described in the DFD of the Business System Option, but other, more powerful tools are used to design the data and procedures.

Elementary Process Description

Variant:	Current Physical

Process ID/Common processing ref:	4.3

Process name:	Update Wallchart

Common processing cross-reference	N/A

Description

The Bookings Clerk receives no. of places booked for a given Course Run, and records on the wallchart the numbers of provisional/confirmed numbers.

After making the update to the wallchart, the Bookings Clerk posts all the delegate details to the Bookings Folder.

Figure 9.19

I/O Descriptions

Variant: | Required System |

Project/System SS plc	Author	Date	Version	Status	Page of

From	To	Data flow name	Data content	Comments
D	7.1	Notice of Cancellation	Course Code Course Date Delegate No. Branch No. Manager ID	Input to process
7.1	7.2	Cancellation Notice	Course Code Course Date Delegate No. Branch No.	
7.2	M S	Old Delegate Details	Course Code Course Date Delegate No.	
7.1	7.3	Request for Standby	Course Code Course Date Branch No. No. of places	
7.3	D	Offer	Course Code Course Date New Delegate No. New Delegate Name New Delegate Address	
M3	7.3	Candidate Details	Branch No. Branch Address Delegate No. Delegate Name	

Figure 9.20

External Entity Description

Variant: *Current Physical*

Project/System SS plc	Author	Date	Version	Status	Page of

ID	Name	Description
A	Branch	A regional unit of SS plc, that places requests, via nominating managers, for training for its members.
B	Tutor	An employee of SS (Training) plc, who delivers sessions on Scheduled Courses.
C	Course Manager	The owner of a Course Title, who is responsible for providing details of courses, and making the decision to cancel Scheduled Courses.

Figure 9.21

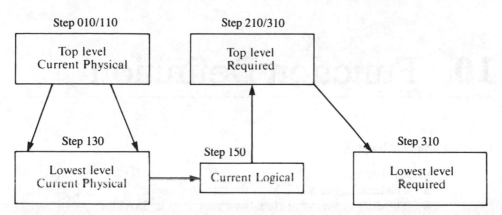

Figure 9.22

EXERCISE
Below is an environment description of an information system. Draw a top-level DFD of the system.

Environment
Old Krate's is a travel agency, specializing in exotic holidays. It holds lists of hotels and charter flights, and creates bespoke holidays for clients. Bookings are made either through a list of OK's agents, or by direct approach from clients.

Procedures
When a client/agent makes an approach, the reservations clerk selects appropriate flight details and hotel details for the customer and makes a provisional booking. The details are entered onto a Provisional Booking file.

The customer must confirm this booking within three days, by sending a deposit of 10 per cent of costs. On receipt of this deposit, Reservations close the Provisional Booking and add the details to their Full Booking file.

Four weeks before the flight is due, Accounts send an Invoice to the client for the balance. Accounts notify Customer Services when the balance is received, and Customer Services then send tickets and joining instructions.

Reminders are sent to customers three weeks and one week before departure. Although the company insists that payment is made at least one week before departure, it has been known that payment has been made and tickets received on the morning of a flight.

At the end of each month, commission of 15 per cent is paid to any agents responsible for holidays commencing during that month.

10. Function Definition

10.1 Aims of chapter

In this chapter you will learn:
- The meaning in SSADM of the term *function*
- The different ways of classifying functions in SSADM
- The composition of Function Definition
- How Function Definition is achieved in SSADM
- The supporting documentation for Function Definition
- The Universal Function Model
- How Function Definition is related to other SSADM techniques.

10.2 Where Function Definition is used in SSADM

Step 330   The different elements for this technique are gathered and identified after the creation of Required System DFDs, in Step 330. We identify the first functions from this model. For each function that we identify, we produce I/O Structures, which will be the input to Relational Data Analysis in Step 340. A User Role/Function Matrix is produced to identify which Users are responsible for each function.

Step 350   The User Role/Function Matrix created in Step 330 is used as input to prototyping the I/O interface. The function entries on the matrix are a result of the function identification described above.

Step 360   The Function Definitions described in Step 330 are amended and increased by the activity of Entity/Event Analysis in this step. As new events are identified, the Function Definitions are amended, and the related activities of Steps 330 and 340 are carried out for them. In Step 360, we develop Enquiry Access Paths for Enquiry Functions and those parts of update processes that are enquiry. The inputs to this activity are the I/O Structures.

10.3 Functions in SSADM

In SSADM the term 'function' applies only to the new system, not the current environment. A function is defined as: 'A distinct piece of the processing carried out in the new system, as perceived by the User.'

The important part of this definition is that it describes the *User's* view of the system, not the analyst's or programmer's. Functions play a significant part in analysis by drawing together all the products from SSADM that define such a piece of processing.

All the functions identified in this stage are carried forward to Stage 5 for detailed Logical Process Design.

Inputs to Function Definition are as follows:

- Logical Data Store/Entity Cross-reference
- Elementary Process Descriptions
- I/O Descriptions
- Required System Data Flow Diagrams
- Requirements Catalogue
- User Roles

Outputs from Function Definition are as follows:

- I/O Structures
- Function Definitions
- Requirements Catalogue
- User Role/Function Matrix

Figure 10.1 illustrates Function Definition in SSADM.

Classification of functions

Functions can be recognized as belonging to one of three types:

1. *Process type*   Update or enquiry, i.e. an update to the database takes place, or data is read without change, respectively.
2. *Implementation type*   On-line or off-line. Many functions are clearly one or the other, but many can be input and processed in either mode. If a function is

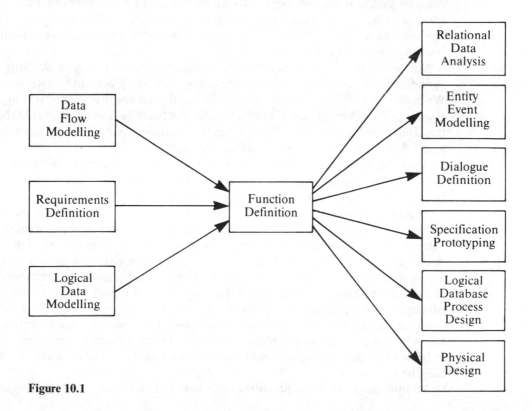

Figure 10.1

implemented either way, a separate function for each should be defined, with cross-reference between them.

If a function can be initiated by an on-line input, but is then processed later, off-line, make a choice as to how it will be classified, one or the other.

3.   *Initiation type*   Is the function initiated by a deliberate input from the User, or does the system automatically trigger its execution? A system-triggered function could be because of time, perhaps, say month-end or quarter-end. Alternatively, it could be because of a condition in the data, such as a customer dangerously exceeding a credit account and cautionary action being taken.

10.4   Procedures

Identify functions

The essential inputs to the initial Function Identification are the DFM of the new system, the associated I/O Descriptions, the LDM of the new systems and the Requirements Catalogue. As stated in sec. 10.2, the identification takes place in Step 330 in the first instance.

Look at the DFD for the required system. Look for all the original triggers of processing, i.e. flows into the system from an external entity, or a process triggered by a data flow from a store (a time-based event).

Once you have identified these, follow all the data flows and processing that belong to each trigger. All those that can be packaged together compose a function. In a complex system it may not be easy to see the dividing line between one function and another, when an input can trigger a succession of different functions; in SS plc, for example, we see the activities that follow the data flow from a delegate cancelling a booking, as shown in Fig. 10.2.

Not all of these activities form one function, Cancel Delegate Booking. Also included are aspects of invoicing, maintaining the waiting list and making bookings. We cannot say ourselves where one function ends and another begins: it is up to the User to tell us where the line is drawn for any one function. Let us say that in this case the User includes Processes 7.1 and 7.2 in the function Cancel Delegate Booking, and that 7.3, 7.4, and 7.5 belong to separate functions.

It may be that Process 7.2 is performed subsequently in batch mode, in which case we would have to create a temporary data store to provide a trigger for that process. If that were the case, the DFD would look like Fig. 10.3.

Perhaps a function is triggered not by time or by an external flow, but by the combination of such a flow with a particular condition on the database: a prospective delegate on the Waiting List from the same unit will trigger another Booking. More drastically, the number of delegates for a scheduled course may fall below an acceptable number and so the Cancel Course process is triggered. In practice, what is more likely to happen in this case is that an exception report will go to the Course Manager, who will make that decision.

So we see that in following the life of an input data flow, we must classify each of the subsequent processing activities according to one of our criteria above before deciding if we are dealing with one or more functions. This activity is described more fully below.

At this point we have identified user-initiated functions, and system-initiated

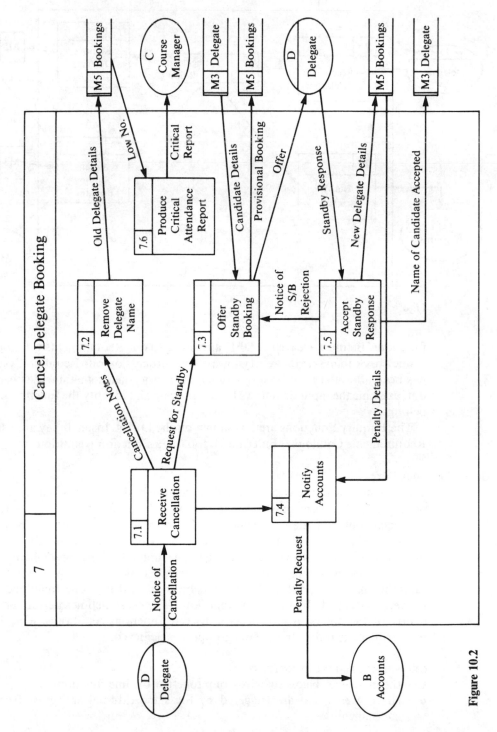

Figure 10.2

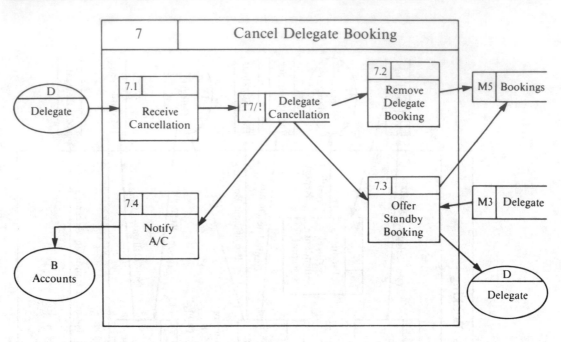

Figure 10.3

functions. Before we leave the DFD and move to our Requirements Catalogue, there is one check to make: that every lowest level process on our Required System DFD has been allocated to at least one function. If not, then something has been missed out; examine the input data flows to that process and identify the function to which it belongs.

The Enquiry Functions are not shown on the DFDs. Instead they are listed in the Requirements Catalogue. Enter these, too, on a Function Definition.

Group functions

We have now, with the User's help, identified the basic functions. That is not the end of the task. The User must now go through all the functions with us to identify sequencing of activities, dependencies between activities, and the possibility of merging some functions. If one function always includes several subordinate functions, then they can be merged into one larger function. Functions 7.3 and 7.5 in Fig. 10.2 could never become just 7.3, as there is a necessary time lag between them. On the other hand, 7.1, 7.2, and 7.4 could be combined into one function, as all are triggered ultimately by the same input flow. If 7.6 is sometimes performed independently of the others, it is kept as a discrete function. As always, it is the User's requirements that dictate how the grouping is achieved.

GROUPING OFF-LINE FUNCTIONS
Use the guidelines below to help group together off-line functions:
- Those events that are triggered by the same external entity, or from related external entities.

- Events which are responsible for outputs to a common destination, or related destinations
- Those events which occur simultaneously, or in close succession.
- Those events which affect common entities.

The rule to follow here is, if in doubt, leave as separate functions. If they do need to be grouped together after all, it can still be done at Physical Design.

The Users in SS plc have decided that the DFD that shows Cancel Delegate Booking is to be defined as the following functions:

1. Receive Delegate Cancellation, comprising 7.1, 7.2, 7.3 and 7.6.
2. Raise Penalty Request, comprising 7.4.
3. Accept Standby Response, comprising 7.5.
4. Produce Critical Attendance Report, comprising 7.6.

Update Functions Definitions Post-ELH

The functions we have already identified are a part of the input to Entity/Event Analysis in Step 360. At least one event is allocated to each function. The normal ratio is one event to one function but, after the grouping exercise, a function may have more than one possible trigger.

As Entity/Event Analysis is so much more rigorous than Data Flow Diagramming (see Chapter 12), it is likely that more events will be discovered than were on the original DFD. Make sure that each of these is allocated to a function. As always, it is the User who will clarify for us how this allocation is to be made, and whether events will be combined for a given function.

Update the Required System DFD and any supporting documentation for events discovered here.

Update Functions Definitions post-Prototyping/Dialogue Design

Possibly the Prototyping activities in Step 350 will result in more functions being identified. If this happens each should be given a Function Definition.

Document the functions

In Step 330, as functions are identified, open a Function Definition Form. Figure 10.4 illustrates a form for the SS plc function Cancel Delegate Booking.

I/O Descriptions referenced on this form are identified by the source and recipient of each flow. Events are identified by the process that receive the input data flow that begins the function processing.

Every subsequent function identified during Steps 350 and 360 must be fed back to 330 for creation of this form.

Draw I/O Structures for each function

Every function, whether update or enquiry, has all the data items on inputs and outputs listed on the Function Definition or Requirements Catalogue. If there are enquiries that have not had these items specified yet, they must be identified at this point. Having identified the individual items for the flow, we must identify extra features, such as iterations of groups of data items, optionally, conditionality, and so on. Any such features must be noted in the comments column of the I/O Structure

Function name	Function ID
Cancel Delegate Bookings	3DG

Function description

This function is occasioned by a Delegate cancelling a Delegate Booking for a given run of a Course Title.

The Delegate details are removed from the Bookings Folder. The Waiting List for that Course Title is examined for other Delegates from the same Branch for ease of billing. Either such a Delegate or else the one waiting longest will be offered a place as Standby.

Error handling

The function is rejected if the system has no record of the Delegate or, if having a record of the Delegate, cannot match it to the Course Run.

DFD processes. 7.1, 7.2, 7.3,

Events: 7.1 Receive Cancellation	Event frequency: 1

I/O Descriptions: D-7.1, 7.1-T7/1, T7/1-m S, T7/1-7.3, 7.3-D, M3-4.3

I/O Structures: 2/1, 2/2

Requirements Catalogue Ref: None

Volumes: Average 15 per week Maximum 30 per week

Related Functions: 4 BI, 5 ST

Enquiries: None

Common Processing: None

Dialogue Names: Delegate Cancellation (DCA)
                Standby Offer (STO)

Service-level requirements:

On-line Response Time	Target: 2s	Range: 2-7s

Figure 10.4

I/O Description

DFD: Required System

From	To	Data Flow	Content	Comments
D	7.1	Notice of cancellation	Course Code Course Date Delegate No. Branch No. Manager/ID	Input to Process
7.1	T7/1	Cancellation Notice	Course Code Course Date Delegate No. Branch No.	
7.2	M5	Old Delegate Details	Course Code Course Date Delegate Duty	
T7/1	7.3	Request for Standby	Course Code Course Date No. of Places	
7.3	D	Offer	Course Code Course Date Del. Duty Del. Name Del. Address	
M3	7.3	Candidate Details	Branch No. Branch Address Del. Duty Del. Name	

Figure 10.5

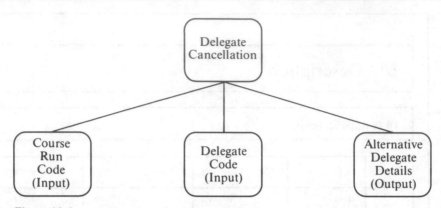

Figure 10.6

Descriptions. Figure 10.5 gives us the description for the I/O to Cancel Delegate Booking.

I/O Structures comprise an I/O Structure Diagram and an I/O Structure Description. The diagram is based on standard structure diagram notation. The data items are represented as boxes, read in sequence from left to right, as in Fig. 10.6.

Those boxes that act as leaves represent data items or groups of data items that cross the system boundary. Each element should be labelled as input or output. Any groups that repeat are shown as iterations (i.e. an asterisk in the top right corner of the box).

If two elements are mutually exclusive, i.e. if when one is present the other cannot be, and vice versa, this is shown as a selection (i.e. a small circle in the top right corner of each alternative box). If an item may or may not be present, that is also shown by a selection, one leaf of which will be null. Figure 10.7 illustrates the notation.

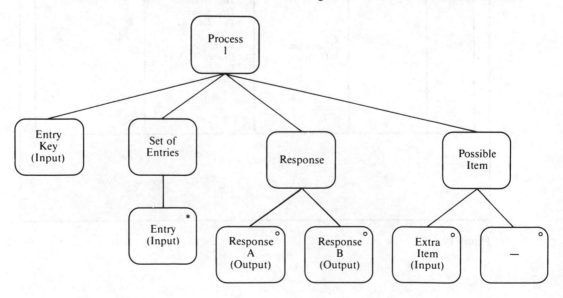

Figure 10.7

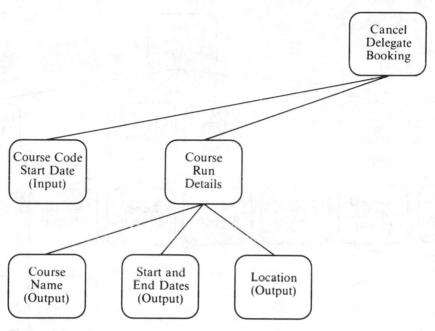

Figure 10.8

OFF-LINE I/O STRUCTURES

For each off-line function, there is one input structure and one output structure.

ON-LINE I/O STRUCTURES

The on-line functions are more complex: each structure is likely to contain both input and output elements.

Let us say that an on-line function, Cancel Delegate Booking, accepts as input the Course Code and Start Date, i.e. the key for Scheduled Course. The systems responds by displaying the full name of the course, start and end dates and the location. This one structure now has input and output elements together, and we have not even executed the function yet! Figure 10.8 shows how this is represented.

So we see that for on-line I/O Structures, we must interleave those elements that are input with those that are output, in order to represent the full dialogue between User and system. These structures are used as the input for Dialogue Design (see Chapter 13).

The full I/O Structure for this function is shown in Fig. 10.9.

10.5 Universal Function Model

Every function in any system can be represented by the model shown in Fig. 10.10. This model is not something created by the practitioner, but rather a schematic, or checklist of all that must be specified in a function.

The model consists of two basic components: processes and *data streams*. Before the function is completely specified, both components must be defined. The definition is built up in pieces, according to the SSADM stage and techniques being applied.

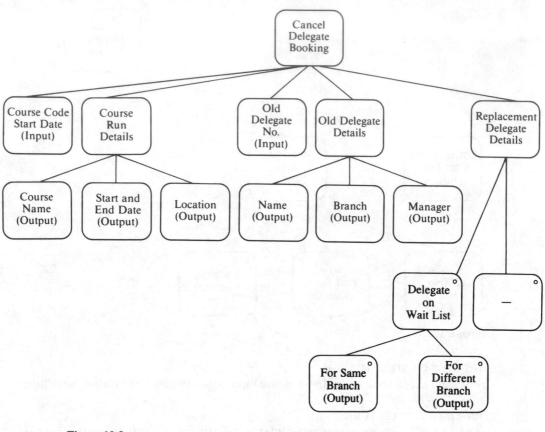

Figure 10.9

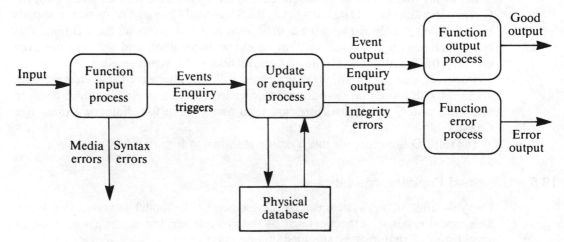

Figure 10.10

Table 10.1

Component	Technique	SSADM product	Stage
Enquiry trigger	LDM	Enquiry Access Path	3
Event	EEM	Effect Correspondence Diagrams	3
Input & good output	FD	I/O Structures	3
Input & good output	DD	Dialogue Structures	5
Event/enquiry output	LDPD	Process Models	5
Integrity errors	LDPD	Process Models	5
Update processes	LDPD	Process Models	5
Enquiry processes	LDPD	Process Models	5
Syntax & control errors	PPS	Program Specifications	6
Input & good output	PPS	Program Specifications	6
I/O process	PPS	Program Specifications	6

Table 10.1 lists the techniques and the appropriate SSADM products against each component.

The Universal Function Model does not present a model of the Function Definition technique; rather it provides a conceptual view of a function.

Data streams

The data streams that flow between the processes are defined in terms of data items only, and represent the logical view. Error handling, integrity error messages, control data, including page numbering, current dates, etc., are not represented on the I/O Structures.

The input data is composed of events, enquiry triggers and redundant data, i.e. data that is input with an enquiry, but is not logically necessary for the requirements to be met.

The data items that make up events and enquiry triggers will be defined during Entity/Event Modelling and Logical Data Modelling.

Processes

The processes identified in the model cover the whole cycle of processing from input, through database processing to output processing. In Function Definition we are concerned only with defining the inputs to and outputs from the function. All the stages of process are defined during the specification of the Process Models in Stage 5, as is the definition of integrity errors.

Syntax errors, control errors, and actual processing of inputs and outputs are left until Stage 6, Physical Process Design.

10.6   Supporting documentation

The documentation for the functions is maintained in the Function Definition. There must be one definition for each function as well as at least one I/O Structure.

The entries in the Definition must record the following information:
● Function ID
● Function Name

- User Roles (refer to the external entities on the DFDs)
- Function Description
- Error handling (only in an informal sense—detailed error-handling procedures are specified elsewhere)
- DFD processes
- I/O Descriptions (refer to those flows across the system boundary)
- Related functions
- Common processing
- Requirements Catalogue Reference
- I/O Structures
- Events
- Access paths (refer to the LDM)
- Enquiry elements
- Volumes (numbers of invocations over a period, peak volumes and times)
- Dialogue names
- Service-level requirements

Summary

Function Definition is one of the central features of SSADM. While it is true that SSADM is a data-oriented method, it is always kept in mind that the data is there to support Functional Requirements, both update and enquiry, on-line and off-line.

We begin identifying functions as soon as we have drawn up the DFD for the Required System, and update the Function Definition with results from Entity/Event Modelling, and from Prototyping.

The Function Definitions are carried forward as input to Relational Data Analysis, Entity/Event Modelling, Specification Prototyping and Dialogue Design.

EXERCISE

Figure 10.11 shows an extract from an elementary process on the Required DFM of Old Krate's. The process describes Make Provisional Booking.

(a) Suggest what candidate functions can be inferred from this process. Classify them, as far as you are able.

(b) From the I/O Description below the DFD, derive and draw the I/O Structure for the function 'Check Hotel Availability'.

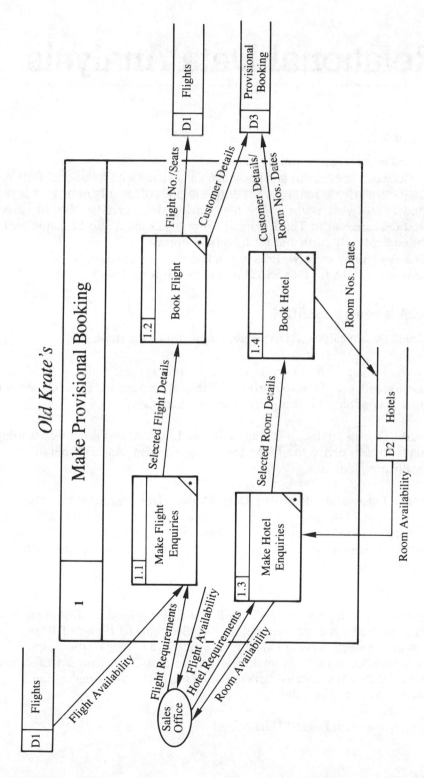

Old Krate's

Make Provisional Booking

Sales Office—Make Hotel Enquires:

Customer Name, Date From, Date to, {Accommodation Type, No. of People},
{Special Requirements}/{Hotel Name, Hotel Address, Hotel Rating}/
Selected Hotel Name
Note: { } denotes repeating groups.

Figure 10.11

11. Relational Data Analysis

11.1 Aims of chapter

By the end of this chapter you should:
- Understand the concepts and vocabulary associated with Relational Data Analysis, particularly the nature of the different kinds of identifying key
- Understand what is meant by normalizing data, and be able to carry out relational analysis to Third Normal Form; you should also be acquainted with the concepts of Fourth and Fifth Normal Forms
- Be able to apply the TNF tests to validate the normalized data
- Know how RDA fits into SSADM, and how it is used with LDM

11.2 Where RDA is used in SSADM

Relational Data Analysis (RDA) is used in the following steps.

Step 140   To help with the analysis of the current system, done—if at all here—in conjunction with the LDM, to help identify the data groups and their relationships. If it is used in this context it is usually in an informal manner.

Step 320   To help produce a Required System LDM. At this point we are using the technique to help derive the LDM for the new system. As in Step 140, it is used informally here, if at all.

Step 340   Enhance the Required Data Model. Here the technique is used more rigorously. The RDA procedures are carried out upon the logical data flows to and from functions (I/O Structures). Data submodels for each major function are drawn up, using the normalized data. These are then used to validate and enhance the Required System LDM.

Use in SSADM

The purpose of RDA is to enhance the Required Data Model. Most of the work in building the model is performed before Step 340, but the rigour of RDA, with its bottom-up approach, serves to identify new entities and new relationships.

The source for data to normalize is the set of I/O Structures from Function Definition. Only a few selected structures will be subject to normalization.

Inputs to RDA are as follows:
- I/O Structures
- Required System Logical Data Model

Outputs from RDA are as follows:
- Set of normalized relations in at least Third Normal Form
- Enhanced Required System Logical Data Model

Figure 11.1 shows the relationship between RDA and other SSADM techniques.

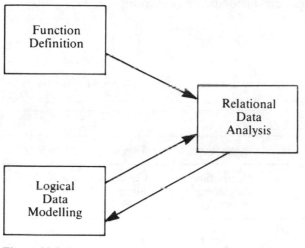

Figure 11.1

11.3   What is RDA?

Like Logical Data Modelling, RDA is concerned with presenting a logical, structured view of data. It is often regarded as being a completely different philosophy from LDM, but SSADM uses it rather as a complement. As used in the method it clarifies and enhances the LDM.

RDA derives from mathematical set theory, and comes from the work of Edgar Codd, of IBM (Date, 1986). Codd, among others, recognized that data in computer systems, and indeed in organizations, was gathered and stored in inefficient ways. This was largely because of the *ad hoc* nature of computerization in the early and middle days of the industry, when computer systems were primarily a way of automating existing manual procedures.

He described a series of steps that formalized and structured data that was previously 'loose', i.e. unstructured, so that data storage was more logical, and independent of specific applications. In this way, update anomalies such as those described in Sec. 11.5 were removed, and data redundancy reduced.

This chapter does not touch on the theoretical underpinning of this method, but is confined to describing how to apply it to any form of loose data. For those who wish to investigate the theoretical aspects in more depth, the works of Codd (1970, 1974), Date (1986) and Fagin (1983) are recommended.

The purpose of RDA is to reduce, if not altogether to eliminate, data redundancy, to resolve update anomalies and ambiguities, to identify all relationships within the data and—in physical terms—to facilitate data maintenance.

The technique involves taking a data source, in SSADM it will be an I/O Structures, or a tangible source such as an input form or screen, and 'normalizing' it (reducing it successively through a series of steps until it is in the most logical grouping) so that the data on it is in a format known as Third Normal Form (TNF or 3NF). To reach this stage we have to take that data through First Normal Form (1NF) and Second Normal Form (2NF).

What we obtain from this process is a notation for a logical representation of data, irrespective of physical implementation and storage.

11.4 Concepts and terms

Relation

A Relation is a term to describe a table of entries. Figure 11.2 shows us an instance of such a relation, with some entries.

Employee

Empno.	Empname	Department	Title	Grade	Mgrno.
A2807	Black	Head-office	Clerk 1	3	C3319
C1166	Jones	Purchasing	Buyer	4	B5834
D4967	Brown	Personnel	Clerk 2	3	A2491

Figure 11.2

The table comprises rows and columns. Each row contains a number of column entries whose sum values make that row unique. Each row is identified by a unique key, or 'true' key. (In Fig. 11.2 the true key is Empno. Empname would not be adequate, as names are not unique to an individual.) The remaining columns describe, or qualify, that key. If we compare this to our Logical Data Model, we can call each relation an entity, and each row an occurrence of that entity.

The rules of relations, as described by Codd in his paper, include the following:
1. Each row must be unique, i.e. duplication of keys is not permitted.
2. There must be no significance in the order of rows.
3. There must be no significance in order of columns.
4. Each column (attribute) may have only one value per row.

Codd has prescribed many more rules to define and limit relations since the earlier papers, but these will serve us for the purposes of this chapter.

Key

A key is a column or group of columns that uniquely identifies a given row in a relation. The terms *true key* or *determinant* are sometimes used: the terms 'key', 'true key' and 'determinant' are synonymous.

SIMPLE KEY
This is a single column that identifies a row. Part Number, Employee Number, and NI Number are typical examples of simple keys to retrieve a given part, employee, or claimant.

COMPOUND KEY
Sometimes a single column is not sufficient to identify a row, and a combination (or

Cust. No:			Order No.

Cust. Name:

Cust. Address:

Cat. No.	Description	Price	Qty

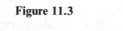

Figure 11.3

concatenation) of columns is required. For example, Fig. 11.3 shows a form for a customer order. We see that it is divided into a header, with an order number and customer details, and entries, which give the details of each stock item ordered.

If we need to access this customer order at any time we need a key. This is easy: every new order that comes into our system is automatically given a unique Order No. To retrieve details of the whole order then, the Order No. is a sufficient key. The individual entries, however, need more information than that to identify them. It may be that one order is for 50 separate stock items. As we are allowed only one value per column entry on a row in a relation, we must treat these order entries differently from the whole order. These entries will form a separate relation (Fig. 11.4).

Cat. No.	Description	Price	Qty

Figure 11.4

To access a given entry on a customer order we must identify a unique key for that entry. We want to identify an entry on a given order, so Order No. would be part of the key; we also want a specific item on that order, so we add the Cat. No., giving us a concatenated, or *compound*, key,

Cat. No. is chosen as the other part of that key, because in this case it is unique to each entry on the order. Other columns such as Description or Qty may not be enough to identify one entry only, without ambiguity or mistake; Cat. No. is sufficient.

COMPOSITE KEY

A composite key, like the compound key, is a concatenated key, but there is a crucial difference. Each element of a compound key is unique: there is only one of each Order No. in our system, and there is only one of each Cat. No. in the system.

One element of a composite key, though, is not unique in the system, and has to be qualified by another element.

In the above example of the Customer Order, the key element inside order is Cat. No. However, each entry on the Customer Order takes up one line on the order form. This gives us another unique identifier for each entry: the Line No. Thus if we were to take the key element as Line No., as in Order 5638/90 Line 1, Line 2, Line 3, etc., we see that the system will have many instances of Line 1, Line 2, and so on.

It is not helpful to us then to ask for details of Line No. 10 in our order system. We will be offered possibly 5000 entries called Line No. 10 to sift through. To identify the one we want uniquely, we must marry it with the order, as in Order No. 5638/90 Line No. 10.

The difference between composite and compound keys is that all elements in compound keys are unique, but in composite keys, one is not unique. In the above example, the compound key is Order No./Cat. No. 5638/90–015699/A. The composite key is Order No./Line No.: 5638/90–10. Either would be sufficient, though, to identify that particular entry on that particular order.

Another example of a composite key is the identifier for a machine on a factory floor. A machine may be a cutter, a drill, a lathe, or whatever. Each *type* of machine will have its own code. For example, a lathe for turning steel may have the code LTST.

In our factory, there may be six such lathes, each one needing to be uniquely identified. As each machine comes into service, it is given a serial number: 1, 2, 3, 4, etc.

'No. 1', 'No. 2', and 'No. 3' by themselves have no meaning in our system: we have a dozen machines with the serial no. 3. What we need is information about the machine that has serial no. 3 and turns steel. In this case, the key would be LTST/3.

CANDIDATE KEY

A candidate key is any possible unique identifier for a relation. One key will be chosen, be it simple, compound, or hierarchic. Others which would serve equally well are candidate keys. They are not in fact keys unless they have been chosen as such. Where a relation has two or more candidate keys, the one selected may be an arbitrary choice.

In SSADM, every candidate key should, at the end of the analysis, be examined to make sure that it is not in fact a determinant. If we take the example of a seating plan for a course, we could find the following relation:

> Course Code
> Course Date
> Room No.
> Seat No.
> Delegate Name

This is an acceptable relation, with a valid key, but it was not the only possible key. An equally valid key would have been:

> Course Code
> Course Date
> Delegate Name

The attributes for this key would then be Room No. and Seat No., giving us an equally valid, and different, relation. Accordingly, we must then include this in our set of relations.

Take care that you do this only with candidate keys that give a new relation, rather than alternative keys that give the same relation, e.g.

> Pay No.
> NI No.

could equally well have been expressed as:

> NI No.
> Pay No.

There is no new information here, and so no new relation.

FOREIGN KEY

A foreign key is a data item in one relation, and a primary or true key in another. The item is referred to as a foreign key in that relation in which it occurs as an attribute. This is a mechanism whereby relationships between data groups can be implemented.

Thus, in Fig. 11.5, an Employee relation, we have the attribute Projno. to show on which project the employee is working. In the same system, we will have a relation Project, whose key is Projno., the same attribute. In the Employee relation, the attribute Projno. is the foreign key, linking a specific employee with a specific project. Conversely, an attribute of Project is Mgrno., to identify the project manager. Further investigation tells us that the value for Mgrno. is taken from the same domain as Empno. Therefore, Mgrno. in Project is also a foreign key, linking the relation to Employee, but in a different relationship.

Employee

Empno.	Surname	Forename	Grade	Projno.*	Start Date
ZO/251	Green	Marion	4	Books90	12/10/90
JO/668	Ashe	Geoffrey	3	Timet90	10/12/90

(Foreign keys are marked with an asterisk.)

Project

Projno.	Name	Start Date	End Date	Mgrno.*
Books90	Bookings	12/10/90	13/06/91	H3/887
Timet90	Timetable	10/12/90	01/04/91	T6/115

Figure 11.5

Dependency

When we say that data item A is *dependent* on data item B, we mean that we need to know the value for B to know the value for A. In the example in Fig. 11.5, Start Date and End Date are dependent on Projno. That is, given a value for Projno., we can find the values for Start Date and End Date.

When a key is identified, the dependency is self-evident; the use of the idea is in identifying the key to begin with, especially when moving to Second and Third Normal Forms.

11.5 How to carry out RDA

Unnormalized form

The first task is to express all data from the source in an *unnormalized* format, i.e. to list all the data items and place them under a suitable (unique) key. This format is termed a *raw relation*.

Let us take a form from Jobs-For-U (JFU). This particular form is a registration

Job-Seeker Registration				No.	
Name:				Age:	
Address:				Date of Birth:	
				Job Code:	
				Alternative Code:	
Tel:					
Qualifications:					
Subject	Level	Year	Result	Awarding Body	

Previous Employment:

Employer Name:	Job Title	Date started	Date left
Employer Address:			
Telephone:			
Reason left:		Pay:	
Employer Name:	Job Title	Date started	Date left
Employer Address:			
Telephone:			
Reason left:		Pay:	
Employer Name:	Job Title	Date started	Date left
Employer Address:			
Telephone:			
Reason left:		Pay:	

Registration Officer:

Duty Code: _____

Name: _____ Telephone no. _____

Figure 11.6

form for new clients, and is filled in by job-seekers when they sign on to JFU's books. It is called a job-seeker's registration form, and is illustrated in Fig. 11.6.

This form presents us with a number of data items about our job-seeker, but these are not in a logical grouping. In fact, the data given to us tells us about rather more than the job-seeker, and so we apply the rules of normalization to give us that more logical grouping.

Unnormalized Form (UNF)

<u>Jsregno</u>
Js-name
Js-address
Js-telno
Js-age
Js-DoB
Js-qualifications:
  Subject
  Level
  Awarding-body
  Year-awarded
  Grade/Class
Previous-employment:
  Employer
  Employer-address
  Employer-telephone
  Job-title
  Date-started
  Date-left
  Reason-left
  Pay
Job-code
Alternative-code
Registering-officer-duty
Registering-officer-name
Registering-officer-telno

Figure 11.7

The first task is to lay out this data as a list of data items, in what is called *Unnormalized Form* (UNF). This comprises the list of data items (attributes, or fields) under a unique, identifying key. Selecting the key in this instance is a simple task, because the form itself provides us with a unique key: Job-seeker Registration Number, or Jsregno. We identify the key field by underlining it. So our first pass at UNF will be as shown in Fig. 11.7. Note:

1. Two sets of data items can have multiple values on this form: those to do with qualifications, and those to do with previous jobs. Those groups of data items are indented under an umbrella name for each. The first umbrella name is Js-qualifications; the second is Previous employment.
2. Each job-seeker is allocated a Job-code to describe the nature of work most suited, e.g. Machine Fitter, Invoicing Clerk, Van Driver, Clothes Shop Assistant, etc. They are also allocated a second such code, called Alternative-code.

Our second pass at UNF involves listing separately those repeating groups and choosing a key for each group. This gives us Fig. 11.8. Each repeating group in this example has a compound key of several elements.

Job-seeker
  Jsregno
  Js-name
  Js-address
  Js-telno
  Js-DoB
  Job-code1
  Alternative-code
  Registering-officer-duty (R/O)
  R/O-name
  R/O-telno

Js-qualifications
  Subject
  Level
  Year-awarded
  Awarding-body
  Grade/Class

Employment-history

  Employer
  Job-title
  Date-started
  Employer-address
  Employer-telno
  Date-left
  Reason-left
  Pay

Figure 11.8

Js-qualifications has a key of Subject (English, Maths, Statistics, Business Studies) and Level (GCSE, A-level, HND, M.Sc. etc). This is because Subject by itself is not unique or sufficiently informative: a Job-seeker may have GCSE Maths, and A-level Maths. The value 'Maths' would not extract all the qualifications. 'Maths' and 'A-level' would.

If an employer were looking for someone with a business studies degree, extracting the Job-seeker and Subject alone would not be sufficient. Excluding Level from the key would give all of the GCSEs as well.

Employment-history has a key of Employer, Job title and Date-started. For the following reasons, the key needs to contain all of these elements to give each entry a unique value. Perhaps a Job-seeker has worked for the one employer on different occasions in different jobs. Employer by itself would not be enough, then, to identify each entry.

Even Employer, Job-title, is not enough: it is possible, after all, that one person has worked as a machine fitter at a plant, left or been laid off, and then returned to the same plant as a machine fitter again, some months later. Therefore that person could have left a job as machine fitter for the one plant twice, giving us two possible values for that data item. Date-started as part of the key will tell us which of those two values is the appropriate one. This gives us Unnormalized Form. Figure 11.13, later in the chapter, shows how this is entered on SSADM forms.

First Normal Form

First Normal Form (1NF) identifies a relationship between groups of data items that corresponds to one-to-many relationships on a Logical Data Structure, as in Fig. 11.9.

The procedure for taking the data from UNF to 1NF is to examine all repeating groups, find their own unique key and make them separate relations.

The key for repeating groups will be a compound key, made up of the key for the original UNF (to preserve the one-to-many relationship) and one or more other elements that will uniquely identify each occurrence of the repeating group.

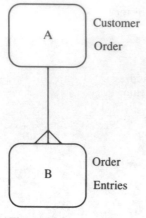

Figure 11.9

Applying this 1NF rule, then, gives us Fig. 11.10.

First Normal Form (1NF)
Job-seeker
<u>Jsregno</u>
Js-name
Js-address
Js-telno
Js-DoB
Job-code1
Alternative-code
Registering-officer-duty (R/O)
R/O-name
R/O-telno
Js-qualifications
<u>Jsregno</u>
<u>Subject</u>
<u>Level</u>
Year-awarded
Awarding-body
Grade/Class
Employment-history
<u>Jsregno</u>
<u>Employer</u>
<u>Job-title</u>
<u>Date-started</u>
Employer-address
Employer-telno
Date-left
Reason-left
Pay
Figure 11.10

Second Normal Form

WHY

The result of putting the data into 1NF means that some data items are identified by a key, without necessarily being dependent on the whole key. Thus, in the new relation, Employment-history, we have the attribute Employer-address dependent on four key items, including Job-title and Jsregno.

This is unacceptable: if we want to remove references to a given Job-seeker, we risk losing information about an Employer on our lists. Conversely, if over 100 of our clients have worked for the one Employer during their history, following mass redundancies from one plant, for example, then for each one of those Job-seekers, we would store the address of the same Employer. To avoid such anomalies, we decompose our relation further to Second Normal Form (2NF).

To achieve 2NF we examine all relations with compound keys for *part key dependencies*, i.e. does each attribute need the *whole* key to identify a value, or only part of the key? In the example just quoted, Employer-address depends only on Employer to identify its value. Date-left, on the other hand, needs the entire key for us to identify a value.

HOW

When going from a 1NF relation to a 2NF relation, copy the entire compound key across first. It may be that one or more data items will depend on the whole key, or it may be that all depend only on part. In either event, the whole compound key must be in a 2NF relation. The key itself has important information connecting, in this case, Job-seeker with Employer with Job-title that must not be lost.

For every other data item in the relation, ask the following question: to obtain one and only one value for this data item, do I need a value for the whole key, or just a part of it?

In the relation Js-qualifications, the answer for each of our data items is: 'We need a value for the *whole* key.' In the relation Employment-history the answer for Employer-address and Employer-telno is that they depend only on a value for Employer, not on Date-started or Jsregno. Therefore it is not yet in Second Normal Form.

All other data items in the relation depend on the whole key for a value, so we leave them as they are.

We can decompose our 1NF relations, then, as shown in Fig. 11.11.

Second Normal Form (2NF)

Job-seeker
  Jsregno
  Js-name
  Js-address
  Js-tel
  Js-DoB
  Job-code1
  Alternative-code
  Registering-officer-duty (R/O)
  R/O-name
  R/O-telno

Js-qualifications
  Jsregno
  Subject
  Level
  Year-awarded
  Awarding-body
  Grade/Class

Employment-history
  Jsregno
  Employer
  Job-title
  Date-started
  Date-left
  Reason-left
  Pay

Employer-details
  Employer
  Employer-address
  Employer-telno

Figure 11.11

If we now remove reference to one of our Job-seekers, we do not lose any information about an Employer's address or telephone number. Conversely, if an Employer moves premises or changes phone number, we need to record that change only once, rather than on every related Job-seeker's record.

Third Normal Form (3NF)

WHY

So far, we have eliminated several storage and update anomalies in our passage to 2NF, but there are still more.

If we return to our relation Job-seeker, we find that on every registration form we hold details of the registering officer. It is right that we identify the officer responsible for registering and classifying each Job-seeker, but details such as name and telephone number do not need to be stored each time and deleted each time we remove reference to a Job-seeker.

HOW

The move to 3NF removes this anomaly by examining dependencies between data items. Thus in our example, we examine the data in the relation Job-seeker, asking for each pair of items: given a value for data item (a), is there one and only one possible value for (b)? Let us look at the Job-seeker relation again to illustrate this. To help us recognize the dependencies, Fig. 11.12 includes a table for JFU employees.

Job-seeker
   Jsregno
   Js-name
   Js-address
   Js-tel
   Js-DoB
   Job-code1
   Alternative-code
   Registering-officer-duty (R/O)
   R/O-name
   R/O-telno

Duty-no	Surname	Forename	Section	Date-started
IN77	Black	Robert	Interview	20.8.87
RE34	Patel	Kaushik	Registry	15.10.86
VA12	Green	Marion	Vacancy	30.5.89
IN50	Patel	Amresh	Interview	12.3.90

Figure 11.12

Given a value RE34 for Duty-no, can we provide one name and one name only? Assuming Duty-no to be a unique key identifying a person, the answer clearly is yes. If we ask the question: 'Given a name Patel, can we provide one Duty-no and one only?' the answer is no; there are two Patels working there, each with a different Duty-no.

Now that we have identified that dependency, we can extract the data items that describe the R/O and assign them to their own key. The actual key of the new relation (in this case R/O-duty) must stay with the original relation, and becomes a foreign key. Thus we can still identify which R/O is responsible for the Job-seeker.

In 3NF, then, we can decompose our relation Job-seeker to the relations in Fig. 11.13.

Job-seeker
   Jsregno
   Js-name
   Js-address
   Js-tel
   Js-DoB
   Job-code1
   Alternative-code
   *R/O-duty
Job-seeker/RO
   R/O-duty
   R/O-name
   R/O-telno

Figure 11.13

On examining all of our relations at 2NF, we see that each data item depends upon the whole key, and only the key, so they are already in 3NF.

Figure 11.14 shows us the complete process from UNF to 3NF. All of this information came only from the single document Job-seeker Registration Form.

UNF	1NF	2NF	3NF
<u>Jsregno.</u>	<u>Jsregno.</u>		<u>Jsregno.</u>
Js-name	Js-name		Js-name
Js-address	Js-address		Js-address
Js-telno.	Js-telno.		Js-telno.
Js-age	Js-age		Js-age
Js-DoB	Js-DoB		Js-DoB
Js-qualifs	Job-code		Job-code
Subject	Alt. Code		Alt. Code
Level	Reg. Officer		*Reg. Duty Code
Award. Body	Duty-code		
Year Award.	Name		<u>Reg. Duty Code</u>
Grade/Class	Telno.		Name
Prev. Empt			Telno.
Employer	**Js-qualifs**		
Emp. Add.	<u>Jsregno.</u>		
Emptelno.	<u>Subject</u>		
Job-title	<u>Level</u>		
Date-start	Award Body		
Date-left	Year Award.		
Reason	Grade/Class		
Pay			
Job-code	**Prev. Empt**	**Prev. Empt**	
Alt.Code	<u>Jsregno.</u>	<u>Jsregno.</u>	
Reg. Officer;	<u>Employer</u>	<u>Employer</u>	
Duty-code	<u>Job-title</u>	<u>Job-title</u>	
Name	<u>Date-start</u>	<u>Date-start</u>	
Telno.	Date-left	Date-Left	
	Reason	Reason	
	Pay	Pay	
	Emp. Add.		
	Emp. Telno.	**Employer**	
		<u>Employer</u>	
		Emp. Add.	
		Emp. Tel.	

Figure 11.14

11.6 Rationalization

The example worked above was taken from just one source document. However, when carrying out the analysis and design of a system, we shall encounter many sources; generally they will be the I/O Structures/Descriptions. Data items will occur several times through the system and this process will identify a number of relations that share the same key.

One of the objectives of RDA is to rationalize data storage and maintenance. The next step is to examine all relations, and merge those that share a key, so that a true key is the key of one relation only. The following steps will achieve this:

1. Examine all the data items for synonyms and homonyms, i.e., in one relation we may have a key Cust-no, and in another, Account-no. Further investigation reveals that both fields describe the customer identifier, but different business functions use different terms. Decide which term you will use and be sure that you always use it in your data model. That is an example of a synonym, two terms describing the one item.

    You may then discover that you have two items called Account-no: one is the customer identifier, as we have seen, and another refers to a financial account held by the company. Here we have a homonym, one term for two discrete items.

2. Having satisfied yourself that all synonyms and homonyms are identified and dealt with, identify all relations that share a key. Merge them so that all of the attributes form one relation.

3. Examine the merged relations to be sure that they are all in 3NF. It is possible that in the merging of relations you have created new inter-data dependencies (or transitive dependencies).

Now you have identified all of the data items in the system, and upon which key each one depends. Having carried out this very detailed analysis of the data, you will convert it into a data model, so that with the LDS you can create the logical model that will provide the basis for the physical database design. This is the subject of Sec. 11.7.

11.7 Enhance Required System LDM

LDM gives us a view of the system data that begins with entities and relationships, and works down to data items via the supporting documentation. With RDA, we begin at the lowest level with the data items and their determinants, and work up to the broader view of entities and relationships. There is a simple five-step algorithm to achieve this.

After we have optimized the relations, and confirmed that they are still in 3NF, we perform the five steps (*note*: the terms 'entity' and 'relation', below, are interchangeable):

1. Represent each relation as an entity, i.e., draw a box and name it after the relation.

2. If the primary key is a composite key, mark the part that is unique, i.e., the 'qualifying' element, as a foreign key.

3. Make sure that all masters of compound key relations are present. This means that every element in a compound key must be the primary key of another entity. If such an element is not the key of another entity, then create one, with that element as the key.

    Any other entity that contains that data element as part of a compound key will also be a detail of this new entity. This element will be marked as a foreign key in all relations where it appears as a non-key data item.

4. Make compound key relations into details. Where an entity has a compound key, it will be a detail of all entities that has one of the key elements as its entire key. This is a simple extension of rule 3.

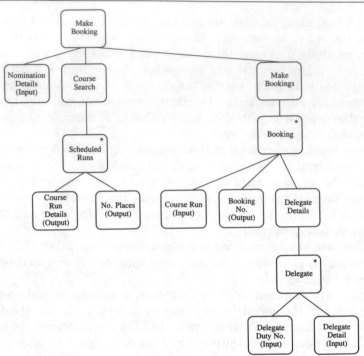

Figure 11.15

5. Make relations with foreign keys into details. Every data item that is marked as a foreign key is, by definition, a primary key of another relation. That other relation will now be shown as a master of the relation with the item marked as a foreign key.

The application of the rules above create another LDM which is similar to the original LDM. It is not intended to supersede the original, but to enhance it. If the functional requirements demand features on the relational model to be included on the data model, then include it. If the new entities forced out by RDA do not serve any functional requirements, do not include them.

11.8 Super Systems plc

This section demonstrates the whole process of normalizing to 3NF, and building a relational model using the five rules. The source for this is the SS plc I/O Structure for the function Make Booking. The structure is shown in Fig. 11.15.

Unnormalized Form appears in Fig. 11.16.

Make Booking

Nominator Duty-code
Nominator Name
Nominator Address
Norminator Phone

Course Title
Course Date
Location
Cost
No. Places
No. Places Spare

Booking No.
Course Title
Course Date
Booking No.
Course Title
Course Date
Delegate Duty-code
Delegate Name
Delegate Phone
Special Factor

Figure 11.16

To reach 1NF we expand the key of the repeating groups, in this case Course Title and Delegate. First Normal Form appears as in Fig. 11.17.

Nominator Duty-code
Nominator Name
Nominator Address
Nominator Phone

Nominator Duty-code
(Course Title)
(Course Date)
Max. Places
No. Places Spare
Location
Cost

Nominator Duty-code   Nominator-Duty-code
(Course Title)   (Course Title)
(Course Date)   (Course Date)
Booking No.   Booking No.
Delegate Duty-code
Delegate Name
Delegate Phone
Special Factor

Figure 11.17

We carry out the part-key interrogation for each relation with a compound key to give us 2NF, as in Fig. 11.18.

Nominator Duty-code
Nominator Name
Nominator Address
Nominator Phone

Nominator Duty-code
(Course Title)
(Course Date)
Booking No.

Nominator Duty-code
(Course Title)
(Course Date)

Nominator Duty-code
(Course Title)
(Course Date)
Booking No.
Delegate Duty

(Course Title)
(Course Date)
Max. Places
No. Places Spare
Location

Delegate Duty
Delegate Name
Delegate Phone
Special Factor

Course Title
Cost

Figure 11.18

Moving to 3NF we examine the 2NF relations for inter-data dependencies. Remember that we are examining not only the non-key data items for dependencies, but also the components of compound keys. The aim is to identify *all* compound keys, to make them determinants in their own right. This can be the most complex part of the RDA procedures, as every set of data items must be examined.

Nominating Officer
Nominator Duty-code
Nominator Name
Nominator Address
Nominator Phone

Booking
Booking No.
*(Course Title)
(Course Date)
*Nominator Duty-code

Course Run
(Course Title)
(Course Date)
Max. Places
No. Places Spare
*Location

Course Title
(Course Title)
Cost

Course Location
(Course Title)
Location
Max. Places

Delegate
Delegate Duty-code
Delegate Name
Delegate Phone
Special Factors

Delegate Booking
Booking No.
Delegate Duty-code
Nominator Duty-code

Figure 11.19

In the example shown in Figure 11.19, we find that after first examining compound key items and non-key items, there is still a determinant hidden in one relation: Course Title may have a different set of Max. Places, depending which location is being used, so another key emerges: Course Title/Location. This is not immediately obvious on the first pass through 2NF relations, but when such a case is identified, a new relation must be created in addition to the existing relation. (This process is known as Boyce-Codd Normal Form. It is a refinement of 3NF rather than an addition to it, and is intended to fill any holes that straightforward application to 3NF procedures might leave.)

We now apply the five conversion rules to the 3NF relations to create the RDA model.

1. Make each 3NF relation into an entity (Fig. 11.20).

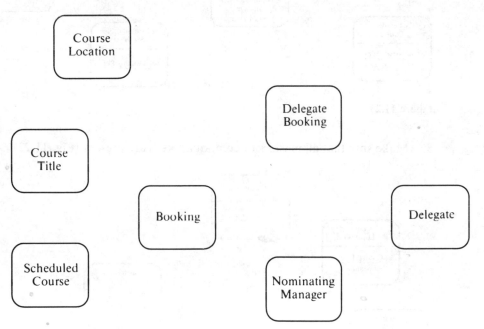

Figure 11.20

2. Mark qualifying elements of composite keys as foreign keys (Fig. 11.21).

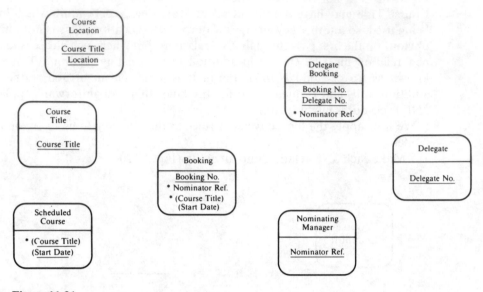

Figure 11.21

3. Make sure that all masters of compound keys are present (Fig. 11.22).

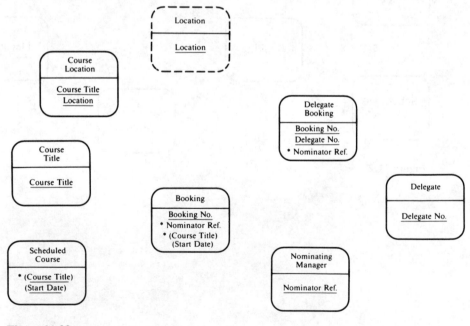

Figure 11.22

4. Make compound key relations into details (Fig. 11.23).

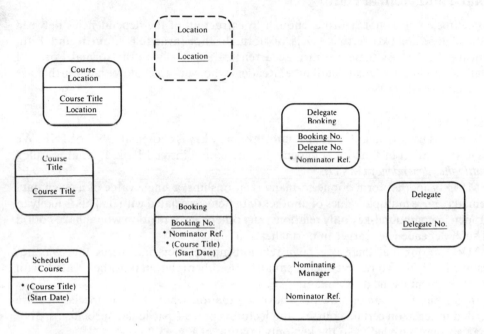

Figure 11.23

5. Make relations with foreign keys into details (Fig. 11.24).

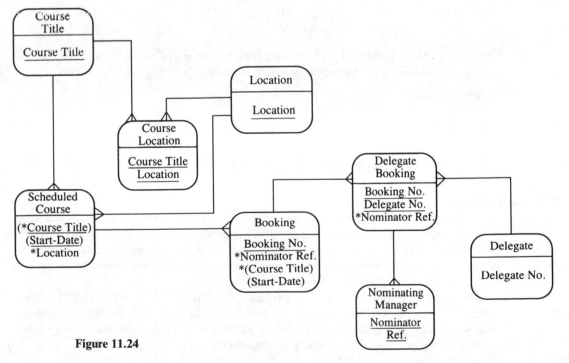

Figure 11.24

11.9   Beyond Third Normal Form

Sometimes 3NF is not rigorous enough to extract all of the dependencies between data. There are two further levels of normalization available, Fourth and Fifth Normal Forms. As these are rare occurrences—especially Fifth Normal Form—I shall not go into too great detail here. Readers who want to explore these further are referred to Date (1986).

Fourth Normal Form

This case is usually found if an unsuitable primary key is chosen at UNF or 1NF. We leave the notion of functional dependency here, and instead look at what is called *multivalue dependency* (MVD).

MVD is another form of one-to-many relationships; a single value of a given data item may have multiple values of another data item associated with it. This is likely to happen in compound-key only relations; any other sort of relation would have got rid of such dependencies earlier in normalization.

MVD means that there is a $1:m$ relationship between data items that was not resolved at 1NF. To resolve such a situation, another relation must be created, with the '1' element as the determinant.

In SS plc, for example, we may have a situation where certain specialisms are needed to teach on certain courses, and lecturers possess particular specialisms. 3NF analysis may have led us to the key-only relation of Fig. 11.25.

<u>Lecturer Code</u>
<u>Course Code</u>
<u>Specialism</u>
Figure 11.25

There are MVDs between Lecturers and Specialisms, and Course Types and Specialisms. To resolve these, we create two new relations, as in Fig. 11.26.

<u>Lecturer Code</u>	<u>Specialism</u>
<u>Specialism</u>	<u>Course Code</u>

Figure 11.26

These are now in Fourth Normal Form (4NF).

Fifth Normal Form

Fifth Normal Form (5NF) has been described by Date as a pathological case because it is so rare that decomposition at this level is needed. As with 4NF, we are dealing with key-only relations, and with keys that contain three or more elements. It may be that they cannot be broken down into two further relations, but can—and therefore must—be broken into three.

There are no examples from SS plc to demonstrate this. If we return to JFU, from Sec. 11.5, we can take an example of a Job-title such as Driver, Fitter, Counter Assistant, etc., being common to several employers. Several Job-seekers may have several different Job-titles at different Employers.

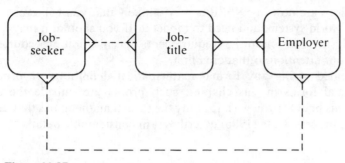

Figure 11.27

The LDS for such a situation is as in Fig. 11.27. The key-only relation reflecting this would be:

<u>Jsregno</u>
<u>Employer</u>
<u>Job-title</u>

To resolve this, without losing information, we have to create three new relations, rather than two. The relations would be:

<u>Jsregno</u>	<u>Job-title</u>	<u>Employer</u>
<u>Job-title</u>	<u>Employer</u>	<u>Jsregno</u>

Figure 11.28 shows the corresponding LDS with these relations.

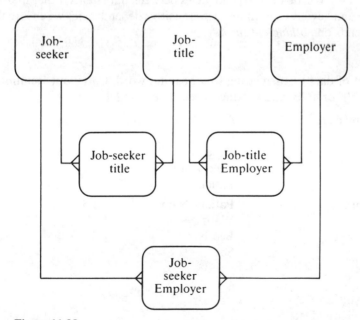

Figure 11.28

These relations are now in 5NF. It is very unlikely that you will ever meet such a case in a real-world system, and need to model it. If you should, it may be a sign of muddled analysis earlier on in the Requirements Specification procedures, and you should turn your attention to that area first.

I have avoided giving any formal mathematical definitions or proofs of the different normal forms in this chapter; such proofs are outside the scope and intentions of this book. If you wish to study the formal mathematics that underlie the normalization process, Date (1986) describes it in considerable detail.

Summary

Relational Data Analysis is intended as a method of identifying the true key of every data item in the system. Using this approach, we expect to reduce data redundancy to a minimum, and produce a model that is flexible and easily maintained.

Because it is based on a mathematical principle, it is rule-based, and so relatively simple to apply, for both novice and experienced analyst.

Another feature of RDA that makes it fit the philosophy of SSADM is its independence from any physical implementation chosen. While it obviously suits a relational database approach, it can equally be implemented on a Codasyl database, or hierarchical or even conventional flat-file structures. Implementation considerations, therefore, neither dictate nor are constrained by the RDA.

- $UNF \rightarrow 1NF$ We identify the key upon which each data item/group depends: by separating out repeating groups we identify for each data item *the key*.
- $1NF \rightarrow 2NF$ We identify which part of compound keys govern each data item: by identifying part-key dependencies we find for each data item *the whole key*.
- $2NF \rightarrow 3NF$ We look for dependencies between data items rather than between data item and key: by examining such transitive dependencies we ensure that each item depends on *nothing but the key*.

EXERCISE

Below is a set of data items relating to a hospital word. Carry out Relational Data Analysis to 3NF or 4NF, and produce a relational model.

Ward Rota (by day)	Consultant's Diary
Ward Name	Consultant
Date	Date
Ward Type	Specialism
No. of beds	Patient No.
Nurse Personnel No.	Patient Name
Nurse Name	Ward Name
Nurse Experience	Bed No.
Patient No.	Sex
Patient Name	
Bed No.	
Admission Date	
Expected Discharge Date	
Ward Sister	

Assumptions: a patient may be moved to a different ward and/or bed during his or her stay. Sister Id. and Nurse No. are the same code.

12. Entity/Event Modelling

12.1 Aims of chapter

In this chapter you will learn:
- The use of Entity Life Histories in SSADM
- The differences between entities, events and effects
- How to construct the Entity/Event Matrix
- How to draw and review the ELH chart
- How to add the operations to the chart
- How to draw supporting Effect Correspondence Diagrams
- How to add the state indicators to the chart

12.2 Where Entity/Event Modelling is used in SSADM

SSADM creates and amends ELHs in the following steps of Analysis and Design:
- *Step 360* Develop Processing Specification
- *Step 520* Define Update Processing Model

Use in SSADM

Entity/Event Analysis presents us with our third view of the data. DFDs and LDSs give us a snapshot of their views of the data; ELHs provide us with the dynamic element of time's action on the system, and so enable us to show (and define) how each entity is updated within the system, whether that update be creation, modification, or deletion.

Inputs to Entity/Event Modelling are as follows:
- Data Catalogue
- Function Definition
- I/O Structure
- Logical Data Store/Entity Cross-reference
- Required System LDM
- Requirements Catalogue

Outputs from Entity/Event Modelling are as follows:
- Entity Life Histories
- Effect Correspondence Diagrams

Figure 12.1 illustrates the relationship of Entity/Event Modelling with other SSADM techniques.

12.3 Entity Life History

ELHs provide us with our third view of the system, the dynamic sequence, or time-based view. It shows the processing cycle of an entity from creation to deletion.

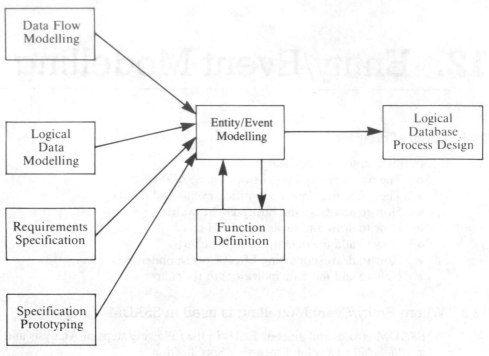

Figure 12.1

It models all the possible changes to the values of the attributes (data items) during its life, and the sequence in which the updates take place.

That may be a little misleading: by 'sequence', I am not describing processing modes, but instead how business rules are implemented. The business rules may stipulate, for example, that an entity occurrence may not be deleted from the system until certain conditions are met. The ELH stipulates the sequence in which updates to an entity must happen, and so is able to build in this particular rule.

ELH uses the structure diagram notation (also known as Jackson-like structures, or Jackson-like diagrams). The structure diagram consists of the three simple constructs: *sequence*, *selection* and *iteration*. Figure 12.2 illustrates a simple structure incorporating these constructs. The sequence is shown by the boxes, from left to right. The selection is shown by the boxes with a little circle in the top right corner. The iteration is depicted by a box with an asterisk in the top right corner.

The single box at the top identifies the entity. The structure is read from left to right, top to bottom, starting with the creation (birth) of the entity and ending with the deletion (death). The boxes at the bottom of each branch are known as *effects*, and show the events in the real world that cause a change to be made to the entity.

A selection shows that at some recognized place in the life history, alternative events could affect the entity. Each possibility is catered for in the structure. Thus, in the example shown in Fig. 12.2, the first thing that can happen after the tutor is appointed will be an induction, or notification that induction will not be required. Perhaps 98 per cent of tutors will need this induction, so we must build in the two

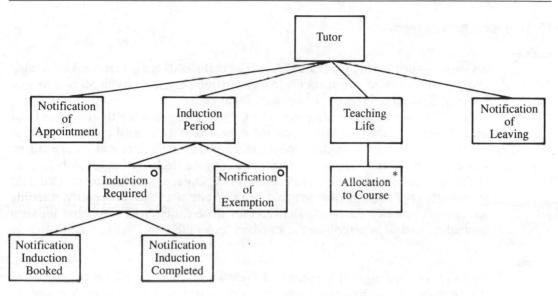

Figure 12.2

events that tell the system of the progress of the induction. For the sake of the other 2 per cent of tutor records, we must build in the exemption, in order to keep integrity and preserve the business rule regarding exemption.

An iteration depicts the possibility that an event may occur several times during the life of an entity, and each occurrence will cause an update to be made. In the example, the repeated event is allocation to the course, and that can only happen *after* the induction. However, a tutor may leave the company before being allocated to any courses; the iteration includes the possibility of that event never happening, so that business requirement is also catered for.

ELHs in analysis

In Stage 3, ELHs provide a validation for DFDs and LDSs, so that we can verify their completeness and correctness. On the one hand we can see how each entity on the LDS is created, modified, and deleted, and on the other we can confirm that the updating processes are connected to the relevant data stores on our DFD.

In Stage 3, the events that affect each identity are recorded, as are the business rules which govern the permitted sequence of activity. Operations are added which explicitly describe the effect of each event, i.e. how the entity is modified, which field, which relationship, etc.

After the ELHs are drawn, an Effect Correspondence Diagram is created for each event. This is carried forward into Logical Database Process Design.

ELH in design

In Stage 5, state indicators are added to the ELHs, to allow semantic validation to take place (i.e. is what is represented on the model a true reflection of what happens in the world?). The Effect Correspondence Diagrams are transformed into Update Process Models, one per event.

12.4   Definition of terms

Entities

An entity, as defined in the LDS and described in the ELH is a generic view of 'a thing about which we hold information'. Delegate is the generic entity in SS plc; Arnold Laine of Solihull is a specific *occurrence* of that entity.

The ELH is a comprehensive picture of all that may happen to any one of our occurrences from the time that the record is created to the time of its deletion. Some occurrences will have straightforward, uncomplicated 'life histories', while others will experience anomalies, e.g. a Customer may be declared bankrupt before an Order is processed, and therefore all details of the Order come to an unexpected halt; another effect of this cancellation could be an update of a Stock Item entity, resetting an attribute of Stock Reserved. The ELH for these entity types must cater for such anomalies, as well as describing the standard 'quiet life'.

Events

An event is something that happens in the real world to cause a change to be made to the database. In our model, it causes updates to one or more entities. We usually, though not always, show the trigger as a data flow across the system boundary. 'Event' does not refer to the processing itself, but that in the real world which causes the processing.

Events are reckoned to be of three kinds:

1. Externally generated, such as a Training Officer placing bookings for staff to attend a course. This will be shown as a flow across the boundary.
2. Internally recognized, such as Course Numbers falling below a critical level, so cancelling a run of a course. This will have an initial trigger of a flow across a boundary, but that itself is not sufficient to be called an internally recognized event; for that, we need the database to be in a particular condition, and so on the DFD we also see another process being triggered by a flow from a data store.
3. Time-based, such as the automatic raising of invoices at a fixed time each month. This will be shown as a flow going from a data store to a process as the sole trigger to that process. Unlike the internally recognized event, though, it will not be traced back to a flow from an external entity.

When naming events, do not copy the name of the DFD process that carries out the update; it may be that more than one trigger can activate that process, each trigger being a different event. By naming an event after a process, a cause of that process may be missed.

RECORD EVENTS

There is no standard repository of event information in SSADM version 4. This is regrettable, but, again, local practice should be able to meet this omission. What is desirable is a description of the events, the entities affected (although of course, ECDs model these), and the volumetrics associated with the event. If a CASE tool is used, it should maintain such details in its own dictionary.

There is implicit information provided, in Function Definitions and Requirements Catalogue entries, but it is a useful idea to be able to refer directly to a form of event definition, as well.

Effects

The changes to an entity are the effects of an event. The leaves of the chart show the events, and must be named after the event.

The effects noted at the leaves are modifications to the database, be that modification the creation of an occurrence, a change to one or more of its fields, or its deletion. An event may also affect an entity by separating it from or connecting it to a master or detail entity. This too must be shown in the diagram.

Sometimes, an event may have different effects on an entity, the effects being mutually exclusive. This can happen, for instance, in SS plc where a delegate is booked on a Course Run. If that delegate is already on the Waiting List for the Course Type, one effect will be to disconnect the Delegate Booking entity from that relationship. If, on the other hand, the Booking is straight from an application on to a Course Run, that will not apply.

Where this may happen, show the possible effects as a selection on the diagram, and in brackets name the possible qualifier. Figure 12.3 illustrates this.

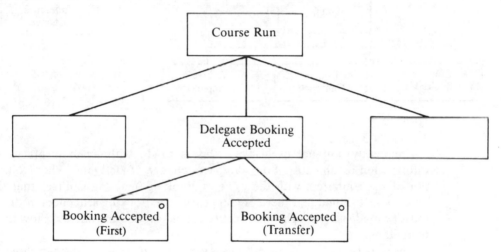

Figure 12.3

Parallel life

The ELH maps out the events that can impact upon an entity in the sequence in which they will occur. A creation event must, by definition, precede a deletion event; an order must be placed before it can be met. However, while most events can be modelled in such a sequence, to reflect the business rules affecting the entity, some will happen at random, without being subject to such rules. A customer can change address while orders are still being processed, for example. The customer record, at least, will need amending, but so will the order record, to show the new delivery address.

Thus, some updates to an entity can occur at any point in the life history; the timing of such events cannot be predicted, and so cannot be put into the strict sequence of events. Another feature is that when the events do occur, they will not impact upon the main sequence of events in the life history.

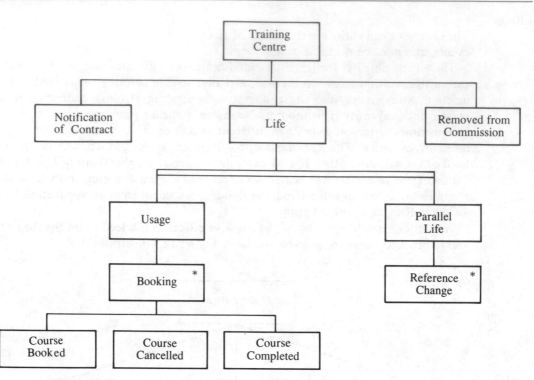

Figure 12.4

Frequently, but not exclusively, this is to do with amendments to reference information on an entity. For example, a change of Delegate's Phone Number will not of itself interfere with the cycle of Booking Training Courses that the entity describes. It should not, therefore, appear in the Booking sequences of that ELH. It must be modelled, though, or the entity could never be amended to show that change happening.

The structure we use to record this is a *parallel life*. The notation to show a parallel life is a horizontal double bar below the 'Life' structure box; from one end of the bar hangs the 'normal life', showing the sequence of events, and from the other hangs the 'parallel life'. Figure 12.4 illustrates this.

There is no limit on the number of parallel lives, or portions of lives, which may occur in one diagram. I have seen one chart with five discrete parallelisms, three of them nested one under another, describing a particularly complicated life history.

12.5   How to derive ELHs

The preceding section is all very well for presenting the theoretical underpinning, but our purpose is to draw, and then use, ELHs. This section describes the procedures for producing the diagrams. As always, I want to emphasize that—like all else in SSADM—this is an iterative process, and it would be less than reasonable to expect to get this exactly right the first time through.

I shall list the sequence of tasks now, and then expand each of them:

- Create the Entity/Event Matrix
- Draw first-cut ELHs
- Review ELHs
- Add operations
- Create Effect Correspondence Diagrams
- Add state indicators

Create the Entity/Event Matrix

As with the drawing of the LDS (Chapter 8), we can draw up a grid to provide us with a working document. Along the top of the grid we list all the entities identified during our investigation. The LDS provides us with this information.

Down the left-hand side of the grid we list all the events that we identify. Our source for this is the set of DFDs already drawn and the associated Function Descriptions.

The lowest level Logical DFDs show us the events. We are looking here for those data flows entering the data stores. By backtracking from these data stores, we can identify the event that caused this update.

SS plc gives us the following example. The extract from the DFD shows us the Course Run data store being updated as in Fig. 12.5.

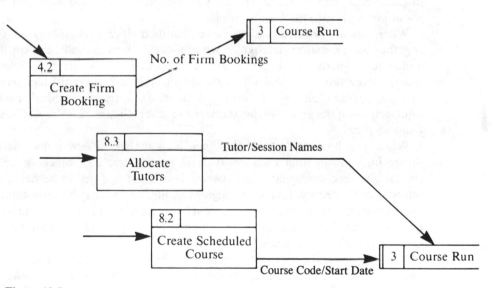

Figure 12.5

Similarly, the input data flow Notice of Cancellation is directly responsible for the modifications to the data stores 1, 4, 3 and 5 as shown in Fig. 12.6

The first attempts at identifying events look at DFDs in this way. Before assuming the grid is complete, however, you should also look closely at the LDM. Identify all the attributes for each entity and, when each is given its value, think about whether that value, like a key, will stay with the entity until deletion or whether it can change. If the latter, have you already found the event(s) that caused that change, or is there another?

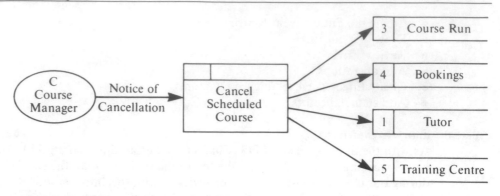

Figure 12.6

Look too at the relationships in the LDS, especially the optional relationships. When are the connections between two occurrences made? What events cause a disconnection/reconnection? Can a detail exist for a period without a master? For example, in SS plc, Session Run is detail to Tutor, but need not be allocated to a Tutor at the time of creation. Has the DFD told you when the connection or disconnection can be made? If not, you may have identified a further event, and should consult with the User about it.

When you are satisfied that you have identified all entities and events, bring them together on the matrix. Recognize what effect each event will have on the entities concerned, whether it is responsible for creating the occurrence, deleting it, or changing the value of a field. Choose a simple code to reflect this, such as C (create), M (modify), D (delete) or I (insert), A (amend), R (remove). Mark each relevant intersection on the grid with the appropriate letter. Figure 12.7 shows a small extract from SS plc.

When you have made your first pass at a matrix, review it by checking each dimension. Ensure that each entity is somehow created, somehow deleted, and preferably somehow updated in between. If it is not updated in between, review the purpose of that entity. It may be a legitimate life, or you may have misunderstood its purpose in your initial analysis. The most likely reason for it not being updated is that it is used for reference purposes only, such as tax tables, and it is another system that is responsible for maintenance.

Next, ensure that each event does actually impinge somehow on the life of at least one entity. If a row or column on the matrix is left clear, you have made a serious mistake somewhere!

It may well appear, at this stage, that the initial analysis has been faulty, and that processes or data are missing from your earlier DFD or LDS documentation. If this turns out to be the case, put them in and reflect the change in the matrix. We often find that ELHs, as a more detailed tool, expand both DFD and LDS by identifying update anomalies in the earlier stages of analysis.

At the same time, you should review the matrix against the entities and Entity Descriptions—perhaps identifying new attributes for update that were missed earlier. If new attributes—or even new entities—are discovered, are there knock-on effects on others when the 'new' one is updated?

Entity / Event	Course Title	Course Run	Booking	Tutor	Location	Delegate Booking	Delegate	Session Run	Room Allocation	Delegate Training History
Booking Received	M	M	C			C	M			
Course Scheduled	M	C			M			C		
Tutor Allocated				M				M		
Location Booked		M			M				C	
Scheduled Course Cancelled	M	D	D	M	M	M		D	D	
Delegate Booking Cancelled		M	M			D	M			
Removal of Delegate		M	M			D	D			D

Figure 12.7

The analyst will not be sitting in a secluded tower pondering these questions; each time such an anomaly is found or query is raised, the analyst must return to the User for clarification. At no point is the poor User regarded as a superfluous agent (or even irritant!) in the process; while SSADM may not be a participative method in the fullest sense, (cf. Mumford's ETHICS, (Mumford, 1983)), the constant consultations make it the next best thing.

It can be seen from the above that although the Entity/Event Matrix is a working document only, it takes considerable time and effort to get it right, and only when it *is* right can we proceed to the actual drawing of the ELH diagrams. To those impatient souls who may ask, is it really worth all the effort and fuss? I can only answer yes! It is *not* worth not taking the effort at this stage: far more time and money will be spent later correcting the mistakes that have arisen from omitting this work.

This exhortation applies to installations that are not using CASE tools. Those using software support should find that in-built integrity-checkers provide the necessary rigour provided by the matrix.

Draw first-cut ELHs

To settle on the 'batting order' for the ELHs, you should have the LDS and the matrix in front of you.

First, using the LDS, select all entities that are details only, i.e. that are not themselves masters of other details. These are the first to be done. The last of all will be those entities that are masters only.

As you begin each entity, you will need the Entity Descriptions to recognize which attributes are associated.

Second, from the matrix identify those events that are responsible for creating and deleting the entity—the first and last events in its life. Draw the creation box.

If any of several events may cause creation of an entity, then represent them all as a selection, as in Fig. 12.8.

Using your knowledge of the system, identify possible sequences in which further updates may happen. We know, for example, that in our entity 'Delegate', a

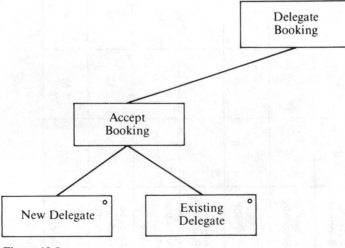

Figure 12.8

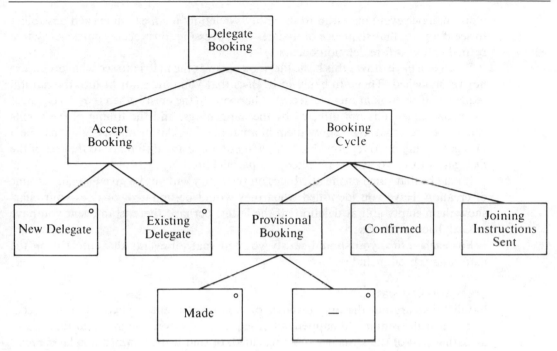

Figure 12.9

Provisional Booking for a Course may be made. Later, any Provisional Booking will be changed to Confirmed. Only later are the Joining Instructions Sent.

In this example, then, knowledge of the system tells us that the update sequence will be:
1. Make Provisional Booking (possible).
2. Confirm Provisional Booking.
3. Send Joining Instructions.
as shown by Fig. 12.9.

It is not so straightforward, of course, because of other events, such as possible Delegate Cancellations, Course Cancellations, and so on, but the illustration should at least make the principle clear.

Third, using our basic structure of sequence, selection, and iteration, draw the simple ELH for your target entity. To help you, although not speed you, you should:
1. *Look at each entity in turn.*   What events are responsible for creating an occurrence of the entity? What events cause an occurrence of the entity to be deleted?
2. *Look at each attribute in turn.*   Identify which events may cause a change to any attribute. Does each attribute have to have a value when the entity is created, or can a value be allocated later? Once a value is given, is it fixed forever, or may it change? If changes are permitted, what event causes them?
3. *Ask*:
    (a) Is it possible to change relationships with a master entity?
    (b) What events occur to establish a change in an optional relationship?

Once you have established a sequence, and confirmed it with your User, and drawn the ELH chart to show it, add the deletion events. As with creation, it may be that more than one event may lead to the entity's deletion; if so, examine each possibility to see if a particular sequence of updates or processing must occur, particularly to a related entity, before deletion occurs.

When you have drawn this basic life, look again at the matrix to see what events are not yet modelled. They are likely to be ones that you could not fit into the natural sequence. If so, look at them to see what happens to the entity when they do occur. If the main sequence is not affected by the happenings, and the timing of the events cannot be predicted, then show them in a parallel life. An instance of this may be a Delegate changing Telephone No. This will not make any difference to the rest of the Delegate's ELH, so will be reflected in a parallel life.

It may be that some of the attributes on the entity will not be given a value at time of creation. Have you identified when they would be updated, or does your chart show them empty still at deletion? If the latter, perhaps there is an event you have missed: back to the User!

For each entity, you should satisfy yourself that all events that affect it on the matrix have been included.

QUITS AND RESUMES

Parallel lives are not the only instance of 'abnormal events'. There are other occasions when an event might happen out of sequence. An event might occur that causes an earlier part of the sequence to be repeated, or that leaps forward to a later event, missing out several parts of the normal life cycle.

An example might be found in the entity Delegate Booking. It may be that after a confirmation of the Booking has been made, the Delegate has to change the Booking to another run of the course, and therefore has to make a Provisional Booking. There is more than one effect here: the entity must change sets, from one occurrence of Course Run to another, and the status of Booking changes from Confirmed to Provisional again.

We use a notation known as *quit and resume* to depict this. By the leaf that shows the departure from sequence, we place a sign Qn, and against the leaf where we recommence the life cycle, place Rn (the n denotes a corresponding numeric character). Figure 12.10 illustrates the quit and resume for the cited example.

The cancellation in this example may not be a request for an alternative, but rather a cancellation pure and simple. To meet this case, the diagram presents a selection, allowing either a rebooking, or a straight cancellation. Thus, the quit will always take place when the request occurs. The importance of this is explained below.

DISCIPLINED AND UNDISCIPLINED QUITS

It is important to recognize that quits and resumes are not a lazy way of not bothering about sequencing. Where possible, build the structure without using quits. However, in the real world, business rules and practice need to cater for abnormal sequencing of effects through some exceptional event.

The SSADM manuals seem to allow quits and resumes to be built in to an Entity Life History willy-nilly; good practice, however, prescribes greater discipline in applying these. There are different schools of thought as to how the discipline should be imposed. At the time of writing SSADM does not incline towards any one path; local standards should govern how any one project or installation uses the

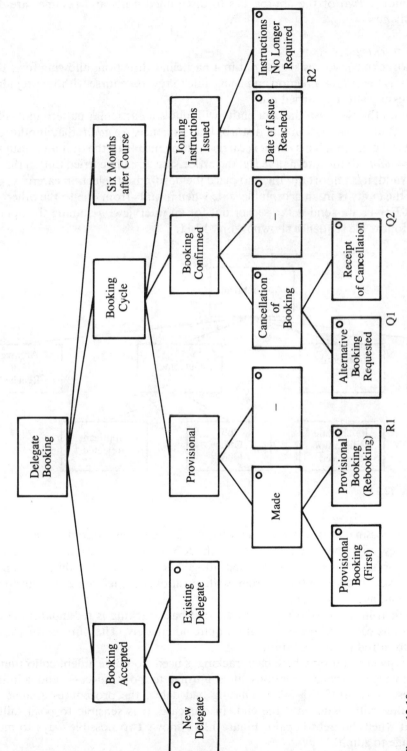

Figure 12.10

technique. Two of the approaches to disciplined quits and resumes are described briefly here.

Backtracking

Backtracking involves setting out in a particular direction, allowing for a discovery that it is the wrong direction and being able to reverse course without prejudicing any processing already carried out.

To do this, we *posit* that our entity will follow a particular pattern in its life, and if we encounter a situation that disturbs that sequence, we *admit* that another sequence holds true instead. This *admit* is carried out by means of the quit and resume. When we do *admit* to the alternative leg, the processing already carried out on that entity is still valid; it is important, then, to place the posit/admit selection carefully, to ensure that the entity is in an acceptable state when it quits from one to the other. Thus, in Jobs-For-U, we send a Job-seeker out for an interview, assuming the appointment will follow the sequence shown in Fig. 12.11.

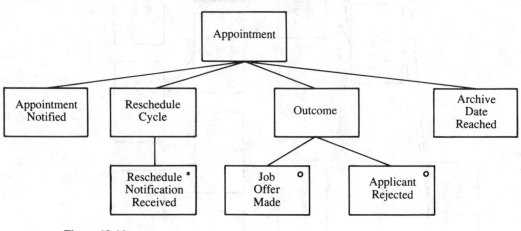

Figure 12.11

It is possible that instead of this sequence occurring without fail every time, the interview might be called off, either by the Job-seeker or the Employer, and so part of that normal life history be omitted altogether. To cater for this, we can make a high-level selection in the diagram, write posit over the left leg and write admit, over the right, as in Fig. 12.12.

In fact, in this case it is easy to see that backtracking is redundant: the ordinary selections cater for the cancellation quite adequately. The purpose of the example was to introduce the notation.

A typical situation where backtracking is used would be a debt collection agency. In most commercial situations, it is normal to posit success, and admit if that success is not fulfilled. When chasing bad debts, the procedures assume that the defaulter will evade until the end. In this case, it is sensible to posit failure, and admit when the debt is paid. Figure 12.13 shows two possible ways to model this passage to admit.

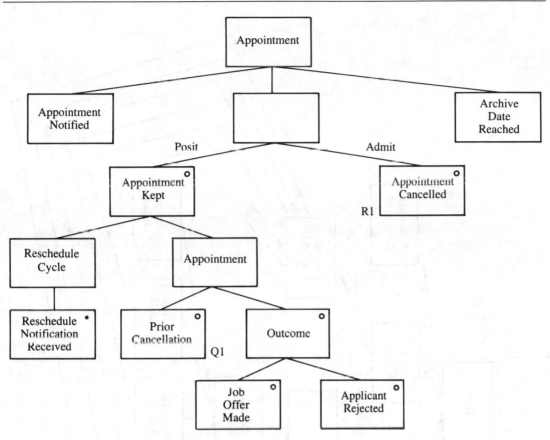

Figure 12.12

Mandatory conditional

If you have to include a quit in your diagram, then the structure should force a quit every time that event occurs rather than allow the sequence to continue sometimes without a quit and sometimes with. Thus, in Fig. 12.12, it would be wrong to model the event such that a cancellation were shown without regard to whether it were indeed a cancellation, or a transfer request. If that were done, then whenever that event occurred, there may be a quit, or then again, there may not. That is undisciplined, haphazard, and suggests that the analyst has not fully understood the situation.

On the first pass, though, do not spend time considering all the unlikely exceptions and sudden deaths in the life history. That will come with subsequent passes.

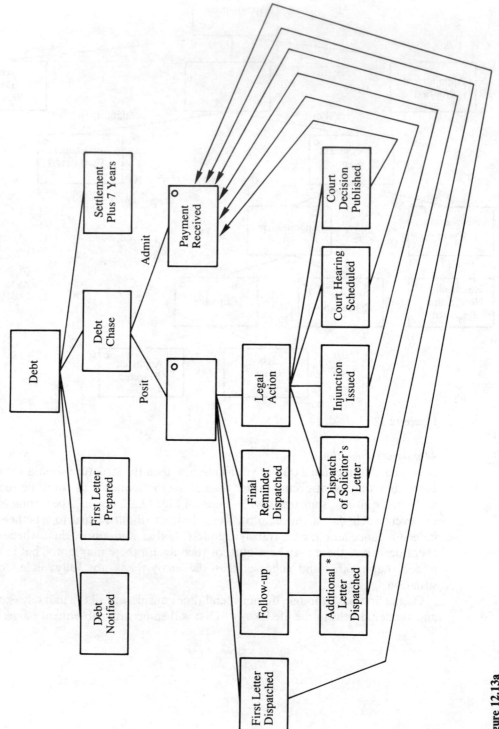

Figure 12.13a

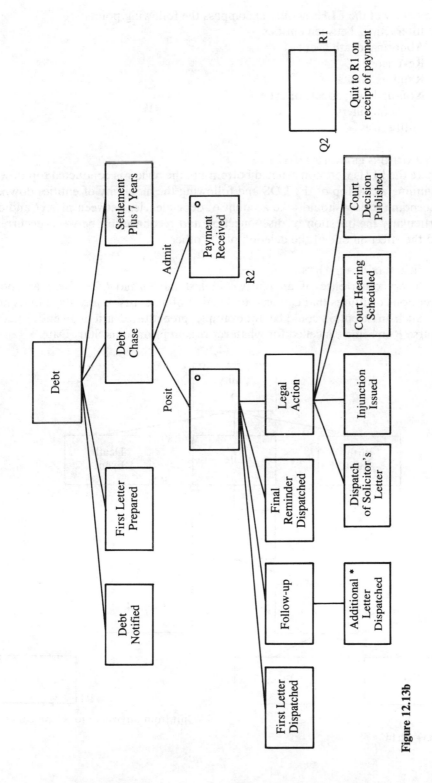

Figure 12.13b

Review ELHs

The review of the ELHs should encompass the following points:
- Interactions between entities
- Abnormal death events
- Reversions
- Random events
- Non-updating effects of events
- Effect qualifiers
- Entity roles

INTERACTIONS BETWEEN ENTITIES

Where the ELHs were constructed bottom-up, the review is conducted top-down, i.e. beginning at the top of the LDS and following the hierarchy of entities down. This descending review should take account of dependencies between master and detail, particularly the question of disconnection and reconnection between occurrences, and the effect on one of the deletion of the other.

ABNORMAL DEATH EVENTS

Not every occurrence of an entity will last its normal life. There are possible exception conditions that can lead to deletion of an entity at any stage in its normal life. Such an exception could be, for example, premature death of an entity, such as a Course Run being cancelled for whatever reason prior to its Start Date.

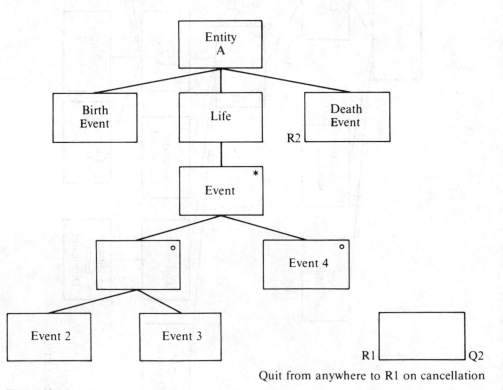

Quit from anywhere to R1 on cancellation

Figure 12.14

This particular exception is catered for by a special use of the quit and resume (see Fig. 12.13b). As this is an event which is unpredictable, and which is not allowed for in the normal life, an obvious and uncluttered notation is called for. This takes the form of a free-standing box at the bottom right corner of the diagram, with a 'resume *n*' symbol by it. Rather than pepper the diagram with a 'quit *n*' at each leaf, the free-standing box implies that the quit symbol applies universally. To reinforce this, a statement to that effect is placed at the foot of the diagram, as in Fig. 12.14.

On premature death, there are three checks to be made before the entity may be deleted:

1. The death of the master is not allowed until all details are first dead.
2. The death of the master kills all remaining details, or
3. The death of the master has no effect on the details.

If either of the first two applies, the death of the master must be carried down to all the details.

REVERSIONS

It may be that from a quit point the entity life will begin again, as in the case of a cheque that has been stopped: after the stop, its life may recommence. In that case, there may be a substructure built in, or a new quit and resume from the free-standing box to a selection under the create box, as in Fig. 12.15.

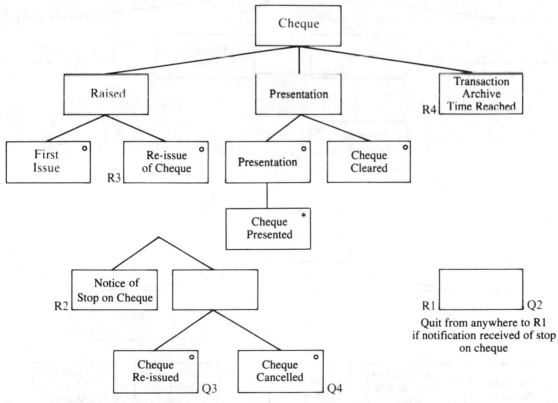

Figure 12.15

RANDOM EVENTS

These are the events whose sequence cannot be predicted, and so are represented on a parallel life.

NON-UPDATING EFFECTS OF EVENTS

It may be that when an event occurs, the state of an entity influences the effect. Thus, if an application is received for Course Run, we would expect to see an effect on that entity, namely an updating of the Numbers Booked. If the Course Run is fully booked already, that event will not affect the entity. It should nevertheless be shown on the Entity Life History.

EFFECT QUALIFIERS

An event can have more than one effect upon an entity. In an environment where payments are made by Instalment, the last Payment will be treated differently from the others, even though the event is the same, ostensibly: Payment data flow crosses the system boundary. To show the difference, the event is drawn twice on the ELH, and the qualifier is marked in brackets below the event name, as in Fig. 12.16.

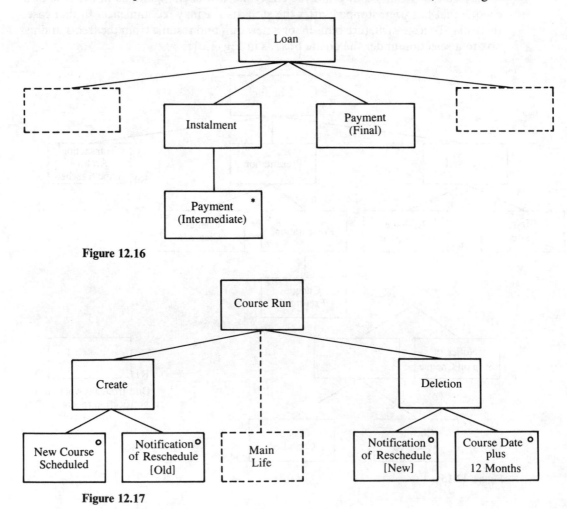

Figure 12.16

Figure 12.17

ENTITY ROLES

Another situation that occurs in the real world is that an event can affect two occurrences of an entity in different ways, depending on their states. An example here might be the Rescheduling of a Course Run. As that would affect the key (Course No./Date), it would cause the deletion of the original Course Run occurrence, and the creation of the new. The same event must be shown on the ELH, and the role of the entity at the time the event happens. Figure 12.17 illustrates this with the portions of the ELH affected. To differentiate this from effect qualifiers, use square brackets. This is also known as a *simultaneous effect*.

Add operations

At each leaf of the ELH, we carry out some operation, whether creating a record, deleting a record, or changing the value of a field. The following operations are used in Entity Life Histories. SSADM does not prescribe what should be contained in ⟨expression⟩. It will usually be a form of simple mathematical operation.

Operation	Description
Store ⟨attribute⟩	Set initial value of attribute on creation.
Store keys	Set the value of the primary key, on creation.
Store remaining attributes	Set the values of remaining attributes as input on creation.
Replace ⟨attribute⟩	Change the values of the attribute to the new value input.
Store ⟨attribute⟩ using ⟨expression⟩	Set initial value of an attribute according to a value in the expression.
Replace ⟨attribute⟩ using ⟨expression⟩	Change the value of the attribute according to the value in the expression.
Tie to ⟨entity⟩	Establish a relationship between target entity and a master entity.
Cut from ⟨entity⟩	Remove the relationship between target entity and a master entity.
Gain ⟨entity⟩	Establish a relationship between target entity and a detail entity.
Lose ⟨entity⟩	Remove the relationship between this entity and a detail entity.

Operations to do with validation or navigation do not belong on an ELH. Include only those which cause an update to be made to the entity or its relationships.

The 'tie', 'gain', 'cut' and 'lose' operations are useful as validation guides: a 'lose' operation on the master should have a corresponding 'cut' operation on a detail, and so on.

Number the operations sequentially, and list them underneath the diagram. Then allocate each operation, by its number, to its appropriate place on the ELH. Figure 12.18 shows the ELH for Course Run with its operations listed and allocated.

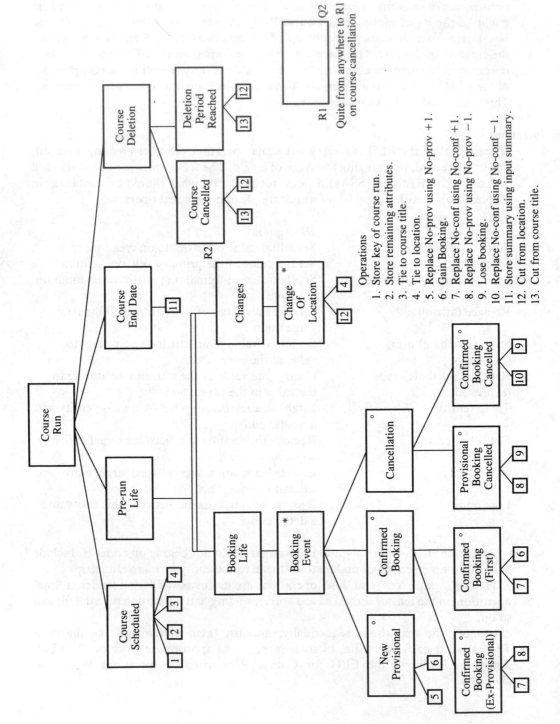

Operations

1. Store key of course run.
2. Store remaining attributes.
3. Tie to course title.
4. Tie to location.
5. Replace No-prov using No-prov + 1.
6. Gain Booking.
7. Replace No-conf using No-conf + 1.
8. Replace No-prov using No-prov − 1.
9. Lose booking.
10. Replace No-conf using No-conf − 1.
11. Store summary using input summary.
12. Cut from location.
13. Cut from course title.

Figure 12.18

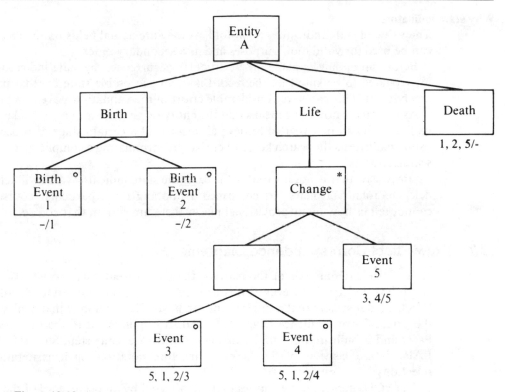

Figure 12.19

12.6 Adding state indicators

What are state indicators?

Expressed most simply, a state indicator is a single-byte attribute that is changed each time the entity is updated. It takes the form of a numeric digit, and is written under each leaf of the ELH, to reflect the values. Each time an event affects the entity, the value of the state indicator is changed to indicate the current state of the records. The example in Fig. 12.19 will make this more clear.

The following rules for adding state indicators should explain Fig. 12.19:

1. Each leaf has two values, a 'before' and an 'after' value. The value is conventionally numeric, starting from 1 at creation, and increasing for each new event that changes it.
2. The first and last values, i.e., on creation and deletion, are null values.
3. On an iteration, the valid previous indicators include the new valid indicator, as in the iteration in Fig. 12.19.
4. If a value is not to be changed, as in a parallel life event, or if any previous value is valid, this is shown as an asterisk.
5. When using a quit and resume, the state indicator is set *before* the quit, so the resume box will show that new value as a valid previous value.

State indicators are added to the diagrams in Step 520, and are the last thing to be done to the ELH.

Why state indicators?

The value of state indicators is twofold: as they are actual fields on the record, they can be used for validation purposes and as a secondary index.

Before an update is made on a particular occurrence, the state indicator can be interrogated to be sure that the record is in an acceptable state for that particular amendment. This can save considerable effort in the validation stage of a program.

As each indicator value means that the entity occurrence is at a particular point in its life, so it can be easier to retrieve all records at a certain stage of processing by using that byte as the search key, rather than by formulating a complicated, multifield search strategy.

Obviously, that is not invariably the case: the state indicator value will tell us that delegate John Matthews has confirmed a booking for a particular course, but it cannot tell us how many provisional places there are still on that course.

12.7   Draw Effect Correspondence Diagrams

As ELHs chart one axis of the matrix—the entity/event axis—so we also need to chart the other axis—the event/entity axis. In other words, we need to identify all the effects on our system that a single event can have. The technique that achieves this is the Effect Correspondence Diagram (ECD). It is equivalent to the Enquiry Access Path, and is built up in a similar manner, during the same step, Step 360. Like the EAPs, it uses aspects of the Jackson structure notation, such as iterations and selections.

The ECDs depict all the entities that are affected by an occurrence of the event in question. Although the *SSADM Manual* (CCTA, 1990) says nothing about the diagram identifying access paths, practice has shown that it is advisable to map the entities on the diagram in such a way that they will reflect the natural order of access when the transaction is processed.

After creation, the ECDs are carried forward to Stage 5 as the basis for the Update Process Models.

How to draw an ECD

The creation of an ECD is straightforward, half of the work already being completed with the drawing of the matrix at the commencement of the step. The steps are demonstrated using the following events: Cancellation of Scheduled Course, Removal of Delegate, and New Course Run Scheduled.

We look at the matrix to identify all the entities that are affected by the event Cancellation of Scheduled Course, and see that they are: Course Run, Course Title, Bookings, Delegate Booking, Delegate, Session Run, Tutor, and Training Centre. The entities Course Title, Tutor, Delegate, and Training Centre are included because they are involved in relationships with those directly affected (deleted), and so will be involved in disconnections from their relationships. We draw the portion of the LDS affected, as in Fig. 12.20. (If you have used a CASE tool instead of a manual matrix it should have the facilities to generate this portion of the LDS automatically. This does depend upon the tool in question, and you should not assume it will deliver like this before you select it.)

Once you have the LDS, remove all the relationship lines and crow's feet, so that

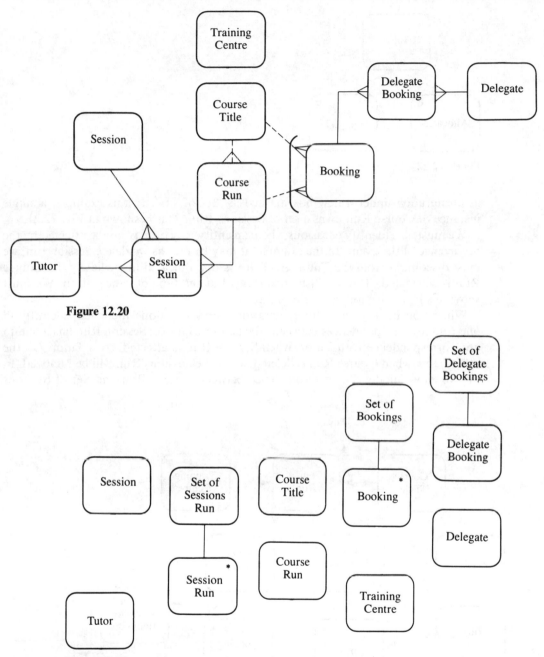

Figure 12.20

Figure 12.21

what you have is a list of entities that you know are affected by the event Cancellation of Scheduled Course.

The next task is to identify whether the entity in question is affected simply, i.e. a single occurrence is created, deleted, or modified, or whether a number of occurrences may be affected, such as all the details of a master. If that is the case we mark it

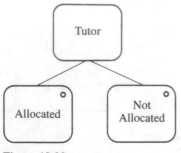

Figure 12.22

as an iteration under a new box marked as 'Set of xxx'. In this example, a single instance of Course Run owns a set of Sessions Run. This is shown in Fig. 12.21.

We must also identify occasions where an entity may or may not be affected by the occurrence of the event. In the example, it may be that as we delete Session run, we must disconnect from the Tutor. As Tutor is not allocated immediately the Course Run is scheduled, it may still be unallocated at the time of cancellation. We must show this by a selection, as in Fig. 12.22.

When you have your entities, iterations, and selections modelled, identify all one-to-one correspondences between effects. For example, Session Run has a one-to one correspondence with Tutor; when Session Run is affected, so is Tutor. On the other hand, when Course Run is deleted, all of the Sessions Run will be removed. In that case, we model the correspondence between Course Run and Set of Sessions

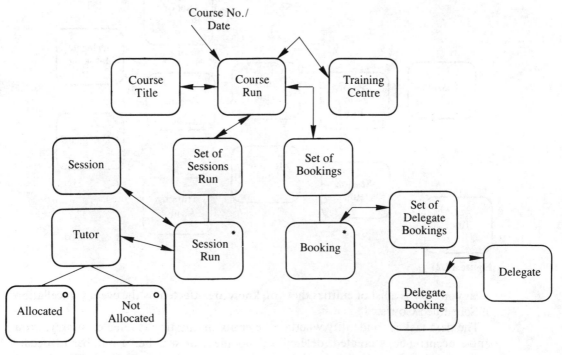

Figure 12.23

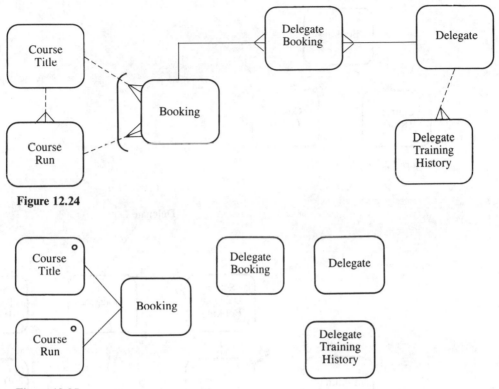

Figure 12.24

Figure 12.25

Run. Show these correspondences as a double-headed arrow between the entities concerned.

Last of all, show the event itself as an input arrow, annotated with the input data, to the entity first accessed. In this case it is the Course Run, with Course No./Date as the input data. This completes our Effect Correspondence Diagram for the chosen event. In full, Cancellation of Course Run looks like Fig. 12.23.

The next event, Removal of Delegate shows two more situations that need to be handled. First of all, the portion of the LDS affected by that event is shown in Fig. 12.24. There is a problem with the Bookings entity: it is the detail of two possible masters, with an exclusive relationship with each. The way to handle this in the ECD is to present a selection between the two relationships, representing the choice as between one of two types of Booking entity, as shown in Fig. 12.25.

Another problem lies with the Booking. Not only is it a choice between Course Run and Course Title, but if it is for Course Run, there is also a choice between a Provisional Booking and a Confirmed Booking. This is an *effect qualifier* identified on the Delegate Booking ELH. In this case, because it will affect only one of the possible Course types, we pass the selection up the hierarchy to Booking rather than to Delegate Booking. This is because it is only at the Booking entity that we identify whether or not the selection should be applied. Both of these features are shown in Fig. 12.26.

An alternative way of drawing the Effect qualifiers has already been indicated:

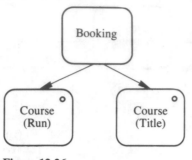

Figure 12.26

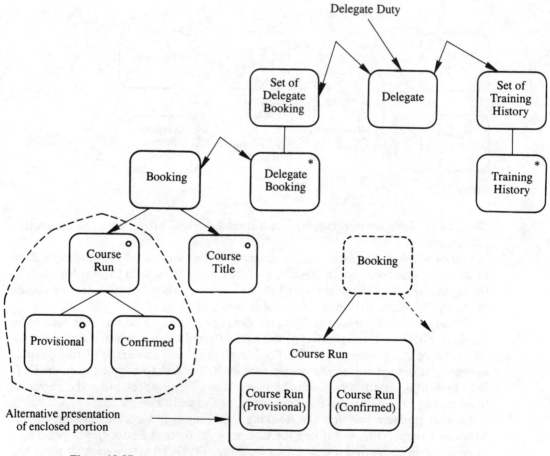

Alternative presentation
of enclosed portion

Figure 12.27

lassoing them together into the one entity box. Whichever method is adopted will be
a matter for local standards.

The full ECD for Removal of Delegates is shown in Fig. 12.27.

Let us now take the event Course Run Scheduled. Figure 12.28 shows the relevant

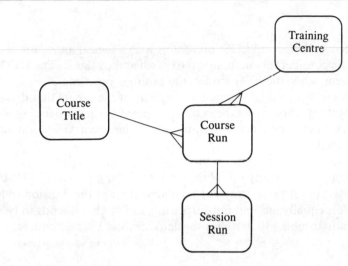

Figure 12.28

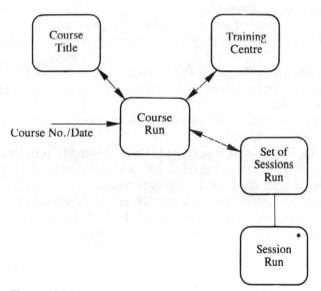

Figure 12.29

part of the Logical Data Model. There are no extra problems involved in this event, so we apply the above principles to produce the completed Effect Correspondence Diagram shown in Fig. 12.29.

The *SSADM Manual* says nothing about the layout of ECDs, whether there is any logical left-to-right sequence of boxes, or whether they are drawn at random on the page. Experience and good practice, however, highlight the advisability of laying out the diagram in the natural sequence, left to right, in which these entities will be accessed. As they will be fed into the process of defining the Logical Update Processing, it makes sense for the navigation path to be implicit in the ECD.

Summary

We can see that whereas DFDs and LDSs present us with a logical and static view of our system, ELHs are considerably more down to earth and dynamic. The DFD and LDS model the system, while the ELH models the business practices.

The ELH provides us with a dual viewpoint of effects of the events that drive our system. The Entity Life History shows the entity viewpoint of the sequence of events that affect it; Effect Correspondence Diagrams show the event viewpoint of the entities that are affected.

The output from this activity is input to the Logical Design activity.

It is a sad commonplace in many establishments professing to use SSADM that because of time/budget constraints, Project Management make the decision to leave out ELH analysis. It is equally sad, but not surprising, to find that it tends to be such projects which run into trouble with both testing activities and User acceptance of the system.

EXERCISES

1. The following environment gives more detailed information on the college described at the end of Chapter 8.

   An entity Loan is created when a Member borrows a book. At any time before Loan expiry, the book may be renewed up to a maximum of three times.

   If anyone has reserved the book, the Loan cannot be renewed.

   If a book has not been returned within one week of Loan expiry the first of three reminders is sent.

   If the book has not been returned within two weeks of the third reminder, the Member is put on a blacklist.

   As soon as the borrower is blacklisted, the Loan entity is deleted and an Outstanding Loan entity created in its place. Otherwise the Loan entity will be deleted six months after the book has been returned.

   The blacklisting will end if the book is returned, or if the price of the book is paid. If neither happens within one year of the blacklisting, the Member's entity is deleted and a Banned record created instead, with the Member's identifier for key.

   Create an Event/Entity Matrix based on this outline. When the matrix is complete, draw the entity Life History for the entity Loan.

   Attributes for Loan are:
   - Book Identifier/Member Identifier ⎫
   - Member Identifier   ⎬ Key
   - Date Loan Created   ⎭
   - Renewal Indicator (values 0, 1, 2, 3)
   - Date Renewal Expires
   - Reminder Indicator (values 0, 1, 2, 3)
   - Reminder Date
   - Title Reserved Indicator

2. Using the information from Exercise 1, draw an Effect Correspondence Diagram for the event Make Loan.

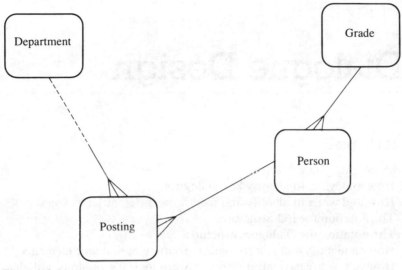

Figure 12.30

3. In Fig. 12.30, there is an extract from the LDS for a Personnel system. The entity Grade will come into being when the first person takes up post in that Grade. There are two ways in which that could happen: an existing employee could be promoted to a new Grade, or a new Person could be appointed to the Company, and the Grade be created for that individual.

   During the Grade's existence, the following events can happen to it: more people can be promoted to that Grade; people can be promoted from the Grade to another; 10 years after a Person has left the company from that Grade, the Person record will be disconnected, and deleted. The Grade record will be deleted 10 years after the last person to hold that Grade has left it.

   Draw the Entity Life History for Grade, including operations.

13. Dialogue Design

13.1 Aims of chapter

In this chapter you will learn:
- How and when to identify key dialogues
- How and when to identify and record the different User Roles
- The notation for I/O Structures
- The notation for Dialogue Structures
- How to identify and record logical grouping of dialogue elements
- How to identify navigation paths between logical groupings of dialogue elements
- How to design Menu and Command Structures

13.2 Where Dialogue Design is used in SSADM

The different elements for this technique are gathered and identified right from the beginning of the analysis, in Step 110.

Step 110—Establish analysis framework   In this step we create the User Catalogue.

Step 120—Investigate and define requirements   The User Catalogue is amended here, in the light of the further investigation.

Step 310—Define required system processing   Yet again, we update the User Catalogue according to our findings and the establishment of User requirements.

Step 330—Derive system functions   At this step, we identify the required dialogues in the new system. Activities include identifying the functions associated with various User Roles, and specifying the I/O interfaces associated with each function. Having identified the dialogues for the new system, we go on to identify those regarded as critical. Mark them on the User Role/Function Matrix, perhaps by circling the relevant intersections. This activity completes the identification of dialogues. The actual design takes place in Stage 5.

Step 420—Select Technical System Options   In this step we create an Application Style Guide, if no such guide exists locally. There is little here in the way of Dialogue Design, but the guide will be used in its creation in Step 510.

Step 510—Design User Dialogues   This is the step when the dialogues are created. The activities involved here are:

Dialogue Design

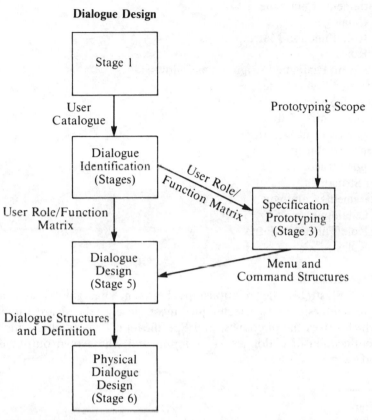

Figure 13.1

- Create Dialogue Control Table
- Create Dialogue Element Descriptions
- Create the Dialogue Structures
- Create the Menu Structures
- Create the Command Structures

All of the above activities will be expanded in this chapter. The sequence of tasks is shown in Fig. 13.1.

Use in SSADM

Although this chapter is entitled 'Dialogue Design', the actual design is the last activity to be carried out. Under this umbrella heading come the activities of Dialogue Identification, Dialogue Specification, as well as Dialogue Design itself.

The specification of dialogues is closely tied in with User Roles, and the identification of these roles is an important part of this exercise.

Inputs to Dialogue Design are as follows:

- Function Definitions
- Installation Style Guide
- I/O Structures
- Menu and Command Structures

- Requirements Catalogue
- User Catalogue
- User Role/Function Matrix
- User Roles

Outputs from Dialogue Design are as follows:

- Application Style Guide
- Command Structures
- Dialogue Control Table
- Dialogue Element Descriptions
- Dialogue-level help
- Dialogue Structures
- Menu Structures
- Requirements Catalogue
- User Catalogue
- User Role/Function Matrix
- User Roles

Relationships with other techniques

Figure 13.2 illustrates the relationships between Dialogue Design and other SSADM techniques. The two techniques most closely related are Function Definition, which drives the dialogues, and Specification Prototyping, which uses the initial identification of dialogues as its input, and whose own output is fed into Dialogue Design.

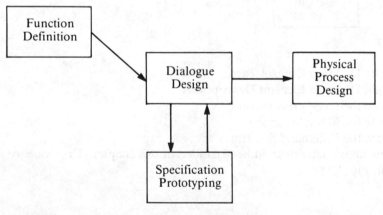

Figure 13.2

13.3   Procedures

Dialogue Identification

PRODUCE USER CATALOGUE

This document is established at the beginning of the analysis, in Step 110. It describes the Users of the planned system—and, if applicable, the current system—by job title and task. Figure 13.3 illustrates the User Catalogue.

User Catalogue

Project/System SS plc	Author	Date	Version	Status	Page of

Job Title	Job Activities Description
Course Manager	To be responsible for contents of Course and, in liaison with Course Scheduling, to arrange for the dates of courses to be run. To authorize cancellation of a course.
Bookings Manager	Responsible for three Bookings Clerks. Team receive Bookings Requests from Nominating Managers to place a number of Delegates on a Course Run, or on a Waiting List for a Course Title.
Bookings Clerk	Receives Bookings request. Places Delegates on Waiting List or Course Run. Updates Wallchart regularly and ad hoc, with Place Numbers. Cancels Delegate Booking. Sends out Joining Instructions
Course Scheduling Manager	Responsible for two Clerks. With Course Managers, arranges Schedule of Courses for three months ahead. Allocates Tutors for each Course and Session.

Figure 13.3

This catalogue is next used to identify the User Roles in the new system, and so it concentrates on two items:
1. The Users in the target population
2. The tasks and functions performed by each User
Note: we do not identify the Users as individuals on the catalogue but as actors of a function. They are identified by job title.

IDENTIFY USER ROLES

By 'User Role' we mean a collection of job holders who carry out tasks in common. The User Roles are identified from the User Catalogue: two pointers that will help you to identify User Roles are:
1. Users with similar job descriptions
2. Users who communicate with the same external entities
Examine the User Catalogue with the above indicators in mind to identify all the User Roles. Some overlaps will be obvious, others will be clear only after analysis or further investigation. Examples in SS plc can be seen in Fig. 13.4.

IDENTIFY THE DIALOGUES REQUIRED

Using the User Catalogue, the User Roles, and the Function Definitions, draw up a User Role/Function Matrix. An example of this is shown in Fig. 13.5, which develops the SS plc User Roles.

Each intersection on the matrix represents a dialogue (where the function is on-line, of course) and so each row identifies all the dialogues a given User requires.

Check the matrix with the Users to make sure that you have indeed identified all the dialogues each wants, and that none are superfluous.

It is quite likely that some dialogues will be optimized, that is to say, if different User Roles are performing the same function, and also use the same data, you can develop just one dialogue for all of them.

IDENTIFY CRITICAL DIALOGUES

Some dialogues may be considered critical to the success of the system, for various reasons. Go back to the matrix, and circle those intersections you identify as critical. By 'critical', I mean that the dialogue meets one or more of the following criteria:
- Dialogues that Users consider critical for their work, perhaps because they define the principal business activities, or are the most used in a given period.
- Dialogues that are shared between many User Roles.
- Dialogues that read from or write to many entities.
- Dialogues that access or input large amounts of data.

With this identification of critical dialogues, the Dialogue Identification activities are complete.

Dialogue Design

IDENTIFY DATA ITEMS

Our input to the design activities is the set of I/O Structures built up from the DFD set and Function Description. Each I/O Structure becomes a Dialogue Structure. Figure 13.6 illustrates the dialogue, Make Booking.

User Roles

Project/System SS plc	Author	Date	Version	Status	Page of

User Role	Job Title	Activities
Booking Clerk	Booking Clerk (customer) Booking Clerk (lecturer)	Make provisional/confirmed bookings for customers or lecturers on courses. Make amendments or delete customer, resource or machine details. Deal with enquiries.
Booking Management	Booking Clerk (management) Bookings Manager	Produce reports. Book, amend, delete any course, resource or machine. Maintain course and resource details.
Invoicing	Booking Clerk (sales) Invoicing staff	Price and post invoices. Maintain price details. Produce statistical reports.
etc.	etc.	etc.

Figure 13.4

User Role/Function Matrix

User Roles \ Functions	Insert New Booking	Delete Cancelled Bookings	Book Rooms For Courses	Replace Cancelled Delegates	Reschedule Delegate Booking	Allocate Tutor	Create New Course		
Course Mgr							X		
Bookings Officer	X	X		X	X				
Scheduling Officer			X			X			

Figure 13.5

Identify the data items for each Dialogue Structure. This information will have been recorded already on the documents supporting the I/O Structure. List each of the items on a SSADM standard form.

IDENTIFY LOGICAL GROUPING OF DIALOGUE ELEMENTS (LGDE)
We will have to navigate between elements of our dialogues. To help design these navigation paths, group the elements in the structure together, in a logical way. Do this in consultation with the User:

● Lasso together the 'leaves' at the bottom of each branch on the I/O Structure, according to the following guidelines. Each set of 'lassoed' elements is an LGDE. Figure 13.7 illustrates this.

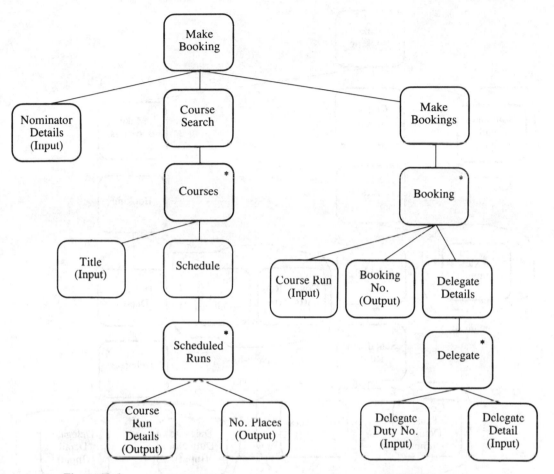

Figure 13.6

- Dialogue elements that occur in a sequence may be grouped together as an LGDE.
- Dialogue elements that are not adjacent or sequential on the structure should not be grouped into one LGDE, unless perhaps the data elements between are included.
- An LGDE must *not* contain only a subsection of a dialogue element. Thus, if the I/O Structure contains the element Customer Details, consisting of <u>Account No.</u>, Name, <u>Address</u>, <u>Phone No.</u>, do not make the LGDE comprise <u>Account No.</u> and Name only. You may have to redraw your I/O Structure at this point, if you want to check the customer's Name only, *before* retrieving the other details. Otherwise, all items are included.
- Generally, an LGDE will embrace both input and corresponding output of a dialogue element. If an input involves a considerable amount of data, then there is a strong case for making the LGDE embrace the input only.

When you have identified the LGDEs and annotated the structure, give each one an ID, numbering sequentially from the left. Figure 13.7 shows the Dialogue Structure for Make Booking with LGDE's identified and annotated.

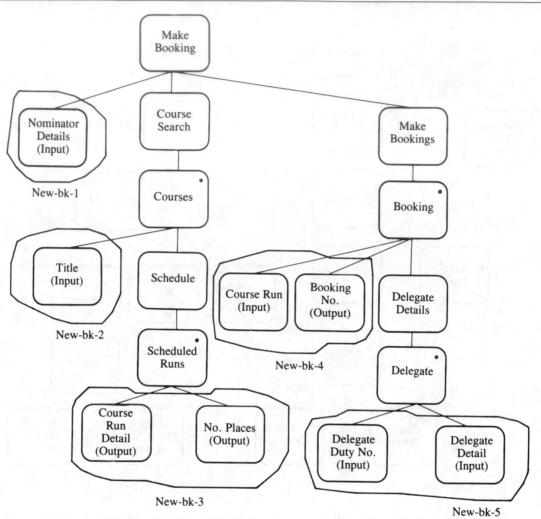

Figure 13.7

IDENTIFY NAVIGATION BETWEEN LGDES

In our Make Bookings example, we have identified five LGDEs. We need to be able to navigate our way between them in order to complete the dialogue. If the Dialogue Structure consists of a simple, uncomplicated sequence, navigation is no problem. If there are null selections and iterations, however, identifying and specifying valid paths from one element to the next is more of a problem.

Figure 13.8 shows this for another dialogue: Delete Delegate. This has six LGDEs, and contains iterations and null selections. Wherever there is an iteration that may have nil occurrences, or a null selection, we recognize the navigation problem.

The SSADM document to manage the navigation is the Dialogue Control Table (Fig. 13.9). This illustrates the possible navigation paths for our Delete Delegate dialogue.

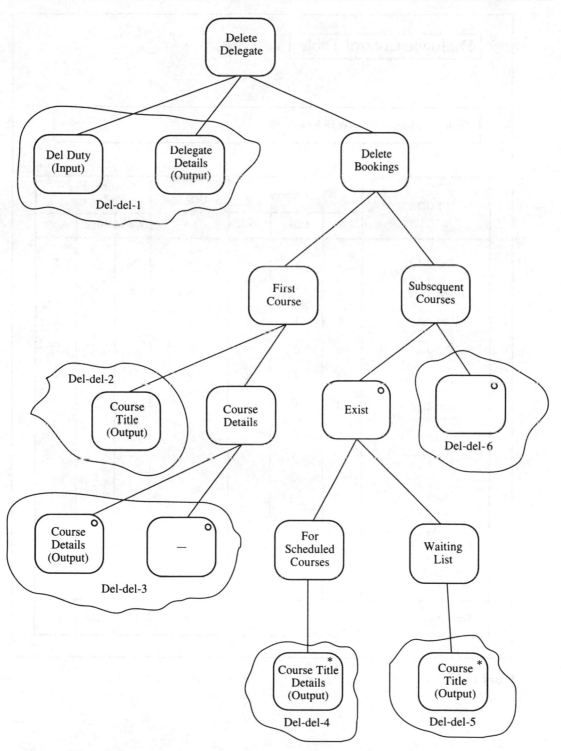

Figure 13.8

Dialogue Control Table

Dialogue name: Delete Delegate

LGDE	Occurrences			Default pathway	Alternative pathways		
	Min.	Max.	Ave.		Alt 1	Alt 2	Alt 3
Del-del-1	1	1	1	X	X	X	X
Del-del-2	1	1	1	X	X	X	X
Del-del-3	0	1	1	X	X		X
Del-del-4	0	20	9	X		X	
Del-del-5	0	5	3	X			X
Del-del-6	0	1	1		X		
Note: the high volume for del-del-4 is to remove all training history records.							
Percentage path usage				70	10	10	10

Figure 13.9

Each LGDE may be classified as mandatory or optional. If the elements succeed one another without possible variation, i.e. there is a simple sequence, the LGDE is mandatory. If the element contains a null selection, or an iteration which may have null occurrences, it is optional. The optionality is documented in a form called the *Dialogue Element Description*. It is used in the Dialogue Control Table.

The first use, then, of this table, is to identify how the system leads the User from beginning to end through the LGDEs, for each time the dialogue is invoked. The fact of optional LGDEs shows that not every one will be used each time, so for design decisions we must identify the most commonly used path through each dialogue. This will be our *default pathway*, which we shall mark on the table.

We may also have alternative pathways that the dialogue can take. Each of these must be entered in a separate column, and an estimate made of the proportion of times that pathway is followed. In the example, there are three possible alternatives; the actual number will depend upon the number of null selections and null iterations within the I/O Structure.

For each LGDE, enter the minimum, maximum and average number of occurrences per dialogue, to help with the later design decisions. The Requirements Catalogue should supply this information.

So we can see that the table identifies alternative paths, and thus enables the designer to create the most efficient set of dialogues for this particular function, based on probable usage.

One last piece of documentation that requires to be completed is the Dialogue Element Description. This identifies all the data items in each data element, and classifies each element as mandatory or optional. Figure 13.10 illustrates the Dialogue Element Description form for the first parts of Delete Delegate.

This is another document that may be produced automatically by a tool, using the I/O Descriptions from the DFM set and the rules of the Structure Diagram.

DESIGN MENU AND COMMAND STRUCTURES

Menus

Menus are a device, usually hierarchical, which give a User or User group access to permitted on-line applications. This design activity is to group together those applications that logically fit together, either by virtue of belonging to the same function, or by being legally available to the same User group.

The SSADM tool we use for constructing menus is the User Role/Function Matrix. Figure 13.11 repeats part of the matrix seen earlier in this chapter. The first thing to do is to identify all of the functions performed by a User Role. The high-level menu could then be looked on as a *superfunction*, which allows Users access to the appropriate subordinate function.

Having identified all the possible dialogues for that User Role, we build them into a hierarchy, by grouping them together in logical order. The actual dialogues identified will form the lowest levels of the hierarchy. As with the LGDEs, there is no fixed algorithm or calculation for deciding on the groupings, so the following rules of thumb will help:
- Bring together in one group those dialogues which logically go together. Refer to the DFD for the required system to help with this; the groupings of the processes

Dialogue Element Description

| Dialogue name: | Delete Delegate |

| User Role: | Bookings Officer |

Dialogue element	Data item	LGDE	Mandatory/ Optional
Del. Duty	Delegate Duty Code		
Delegate Details	Delegate Name Address Branch Telephone No. Training Officer	Del-del-1	M
Course Title	Course Title	Del-del-2	M
Course Details	Title Date, Location	Del-del-3	O
Course Details	Title Date, Location	Del-del-4	O
Course Details	Course Title	Del-del-5	O

Figure 13.10

User Role/Function Matrix

User Roles \ Functions	Insert New Booking	Delete Cancelled Bookings	Book Rooms For Courses	Replace Cancelled Delegates	Reschedule Delegate Booking	Allocate Tutor	Reschedule Course Run	Confirm Booking
Bookings Officer	X	X		X	X			X
Scheduling Officer			X			X	X	

Figure 13.11

on that diagram will give you pointers, as will consultation with the Users, of course. In SS plc, all dialogues to do with Course Bookings could be grouped together on one level: Making Provisional Bookings, Confirming Provisional Bookings, Cancelling Bookings, and so on. This would give a superfunction of Bookings.

Another level would give access to functions regarding Course Run, such as Allocating Tutors, Booking Rooms, Issuing Joining Instructions, and so on. The superfunction here would be Course Run.

- Reflect the User's way of performing the tasks. Our structure should support whatever sequence the User chooses to carry out the task. If this means invoking the one dialogue from different places in the menu hierarchy then that must be permitted and shown.
- Each grouping in the hierarchy may lead either to another lower level menu or to a dialogue on the bottom level.
- There is no need for groupings to have the same number of items: logic, not symmetry, is our aim.

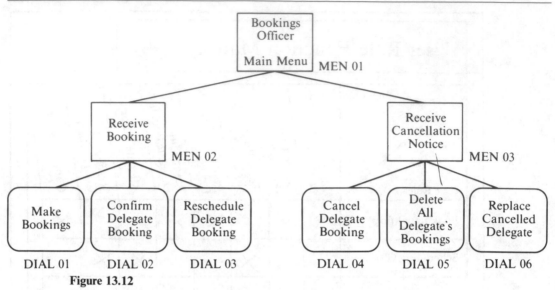

Figure 13.12

Another important input to this activity is the result of any prototyping exercise that involved the use or design of menus.

Having built up our menu hierarchy, we represent it as a tree structure, with the User's entry point to the system the single node at the top. This is shown in Fig. 13.12.

We can see from this diagram that each node may lead either to further menus, or directly into the dialogue. The thing it may *not* do is lead from a dialogue to a lower level menu. The only way from a dialogue is *up* at least one level to a menu. This may be the menu that invoked the dialogue, offering the User another dialogue, or to a higher level, or to a parallel menu.

Each menu node on the diagram is depicted by a square box, referenced MEN *nn*, where *nn* equals a number. Dialogues are depicted by a round-cornered box, referenced DIAL *nn*, where *nn* equals a number.

At the end of the structure for a given User Role, cross-check with the User Role/Function Matrix. Every cross on the User Role's row must be represented by a dialogue box at the bottom of one branch of the hierarchy. Every User Role on the matrix must have one menu structure.

Command Structures

Command Structures show the directions that control can take when a dialogue is completed. This means that dialogues can be invoked with or without menus.

The Command Structure is essentially a form—one for every dialogue—that documents all the possible directions a User can take on completing that dialogue.

Figure 13.13 shows the Command Structure for Cancel Delegate Booking. At the close of the dialogue the User has three courses of action open:

1. To return to the main User menu
2. To continue the same dialogue with another transaction; this means a return to the beginning of the dialogue, but not back to the menu.
3. To quit directly to a related function such as Amend Transfer Fee, or Cancel Delegate.

```
┌─────────────────────────────────────────────────────────────┐
│  ┌──────────────────────────┐                                │
│  │  Command Structure       │                                │
│  └──────────────────────────┘                                │
│                                                               │
│  ┌─────────────────────────────────────────────────────────┐ │
│  │  Dialogue name:    Delete Delegate                      │ │
│  │                                                         │ │
│  │  User Role:        Bookings Office                      │ │
│  └─────────────────────────────────────────────────────────┘ │
```

Option	Dialogue or Menu	Dialogue/Menu name
Input Delete All Bookings	Dialogue	Delete All Delegate Bookings
Quit to Menu	Menu	—
Reschedule Delegate	Dialogue	Reschedule Delegate Booking
Find Replacement	Dialogue	Replace Cancelled Delegate

Figure 13.13

The command structure listing these must contain only items valid for that particular User Role. Refer to the User Role/Function Matrix for confirmation that you have offered all the choices, and only the valid choices.

The actual mechanism for navigating through the dialogues will be left to the Physical Design. Such mechanisms will include the use of function/control 'hot keys', or typed commands to move the User from one part of the menu or dialogue to another.

When making decisions about the Command Structures, take into account such factors as:
- Constraints imposed by selected hardware.
- Relative frequency of dialogues and dependencies between dialogues.

One word of warning when implementing a command structure. If a check for, say, Customer Number; in a dialogue means that a transaction may be aborted near the start of a dialogue, it is the Dialogue Structure that should let the User out, *not* the Command Structure. One project I helped on as a consultant contained a number of dialogues with just such checks, each of which demanded the aborting of the transaction. According to the I/O Structure that was at the base of the dialogue, there was only a low-level selection asking: 'Is Customer No. on DB?' If the answer was no, a subroutine was invoked, which set up the Customer Record, and control then returned to the initial dialogue, rather than picking up the transaction where it had left it. The team had built in a path in the Command Structure that terminated the dialogue in mid-selection to invoke the subroutine, without allowing it to be resumed afterwards.

The situation is a simple and common one, but the team had not thought through the consequences of building the structure in such a way that the User could escape. How this could be addressed may depend on the tool used to design and/or implement the system. If a 4GL is the proposed tool, then the documentation must be annotated to show that this is what will happen. If the implementation is done through a 3GL, though, the I/O Structure must be amended to show that the transaction may be ended at that point, and then the Command Structure may invoke the dialogue afresh, to continue with the ordering process.

Users who are logged on at the terminal will not be aware of how the transactions are being processed, if the messages are worded carefully, but equally, they will not suddenly find themselves in a completely different part of the function or superfunction without warning.

DEFINE DIALOGUE-LEVEL 'HELP'

At this point it may be useful to define the navigation help procedures associated with each dialogue. This is not the place to design help screens, that comes with Physical Design, but to identify the requirements for help facilities that the User may wish.

Help may be wanted on three levels:
1. *Context*   Where am I in the dialogue/menu structure?
2. *Job-related*   What do I do now?
3. *Navigational*   Where's the way out?

If these facilities impact on the associated Dialogue Structures, then the structures must be altered and the supporting documentation revised.

Summary

Dialogue Design is an activity that takes place towards the latter part of the analysis and design cycle, yet preparation begins at the very start in Step 110.

SSADM is concerned with improving the specification and design of on-line systems. The emphasis now placed on the User Roles and Function Definition enables us to carry out this specification more rigorously than before. We can develop the Dialogue Structures from what has been uncovered before, rather than begin a new area of analysis with the attendant risks of missing important requirements previously identified.

Dialogue Design is built up on the I/O Structures developed in Function Definition, by lassoing together data elements on the structure that go together logically. A Dialogue Element Description form defines the contents of the elements, while navigation between elements is described on the Dialogue Control Table.

As an activity, Dialogue Design in SSADM is concerned with defining the logical interface with the Users, rather than considering specific physical issues, such as numbers of lines of a screen, numbers of pages to a logical screen, and so on; these are matters to be resolved during Physical Design.

EXERCISE

Figure 13.14 is an I/O Structure for a function, Cancel Booking, from a theatre company, RunanRun. This is how the function works: cancellations are made by an Agent, who has a unique reference number. After the Agent's identifier is input by the clerk, the Agent Details are displayed, including a list of current Booking reference numbers.

The Bookings may be for a number of Seats on one or more Dates, and are identified by a Book No., which refers to a batch of Bookings made by an Agent. If the Booking No. is not found, the clerk will exit from the dialogue. This happens on about two per cent of occasions.

When a Book No. is input, details of all future bookings associated with that batch are output. The Date, followed by each Seat No. to be cancelled, is entered.

When all seats for a given date are cancelled, Confirmation is displayed. This includes the amount that is to be refunded. The average volume is two seats on a single date being cancelled, with a minimum of one, and a maximum of 20 being allowed.

The one dialogue allows up to five dates to be cancelled. On average, two dates are cancelled within one dialogue.

After the cancellations are made, the operator is asked to identify the method of refund. In 20 per cent of cases, it will be a Credit Note, in 10 per cent of cases, no refund will be applicable, as no bookings have been cancelled.

From the I/O Structure, identify the LGDEs, and then complete the Dialogue Control Table.

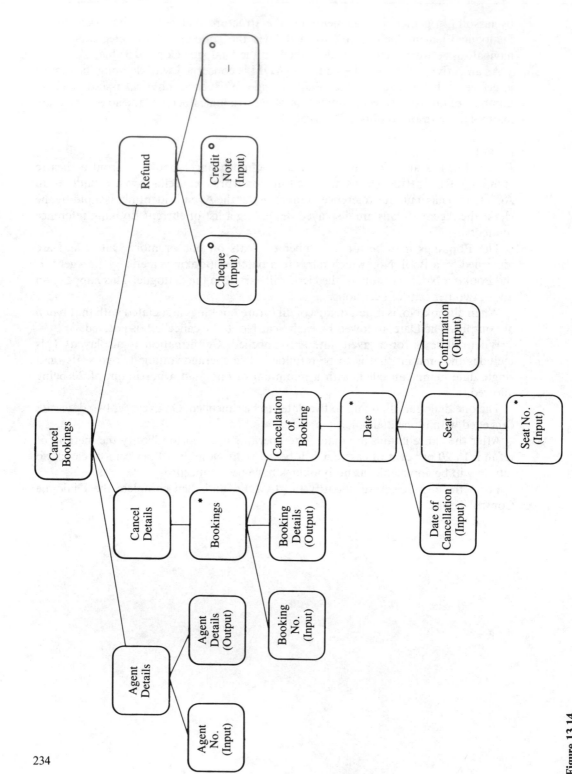

234

Figure 13.14

14. Specification Prototyping

14.1 Aims of chapter

In this chapter you will learn:
- The place of Specification Prototyping in SSADM
- The reasons for carrying out Specification Prototyping
- The inputs to Specification Prototyping
- The products of Specification Prototyping
- The conduct of Specification Prototyping sessions

14.2 Where Specification Prototyping is used in SSADM

Step 350—Develop Specification Prototyping

Use in SSADM

In SSADM, Specification Prototyping has its own prescribed place and procedures. There are other forms of prototyping that can be used in systems development, whether in a SSADM project or another type of project. The other approaches are not described here. Guidelines on the use of these techniques may be found in the Prototyping Interface Guide (referred to in Chapter 2).

Specification Prototyping is a method of examining the workings of the Requirements Specification with the User, with the intention of trapping errors and helping the User to identify new requirements. It is not a method of developing or designing a physical system.

Critical dialogues are the first to be selected for the prototyping process. After, any other dialogues that the User wishes to see trialled can be tested.

Each dialogue chosen is run in a prototyping session, i.e., a computer simulation of the dialogue is played, using dummy data and inputs. Any changes to the dialogue that surface as a result of this are made, and an amended version played. All such changes are documented, as is the Requirements Catalogue.

Inputs to Specification Prototyping are as follows:
- Data Catalogue
- I/O Structures
- Installation Style Guide
- Prototyping scope
- Requirements Catalogue
- Required System LDM
- User Role/Function Matrix

Outputs from Specification Prototyping are as follows:
- Command Structures

- Menu Structures
- Prototyping Report
- Requirements Catalogue
- Prototype Demonstration Objective Documents
- Prototype Pathways
- Prototype Result Logs
- Screen Formats

Figure 14.1 illustrates the relationship between Specification Prototyping and other SSADM techniques.

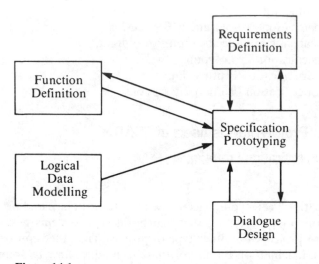

Figure 14.1

14.3   Preparation for the prototyping session

Selection of the tool

As the technical environment may not be known at this point, the target implementation environment is obviously not a candidate for use in the prototyping session. Plenty of tools exist on the market, sharing several essential features: screen painter, data dictionary, and on-line navigation.

Input to the prototyping activities include some SSADM products such as the LDM for the required system. If you are using a CASE tool for the project development, that tool itself may be used for the prototyping; at least, it should be able to interface with the prototyping tool.

Whatever tool is used, it should be chosen and procured early in the project life. If the IS strategy had dictated the technical environment before TSO, then the prototyping tool should be chosen to simulate that environment as closely as possible. If no indication has yet been made as to environment, then obviously that cannot be done.

Scoping the prototyping activities

This activity should be carried out at the start of the whole project. The first thing to do is to identify any need to perform prototyping. If a project has one of the following characteristics, then prototyping will probably not be appropriate:

- The new system is being translated onto the new system directly. If the change is merely an upgrade of hardware/software, there will be no need to trial the Requirements Specification this way.
- The nature or size of the project does not justify the cost of resources for prototyping.

If prototyping is justified, is it screen prototyping, or report output prototyping? Probably both will be required for the system. In any case, you should establish the following before you begin.

SCREEN PROTOTYPING

- What is the likely level of on-line activity? If there is considerable interaction, prototyping will help validate the specification requirements.
- What is the likely level of data manipulation on screen? If for any one function the activity is large, prototyping will assist in validating and collecting the User's needs.
- If the on-line interaction is poorly thought out, will that have a detrimental effect on the business, or will it just be inconvenient at a local and trivial level? If the former is the case, then prototyping will be valuable.

REPORT OUTPUT PROTOTYPING

- If an output from our target system is going to serve as input to another, prototyping can help to ensure that the output meets the requirements.
- If an output has to meet certain statutory requirements, tax return forms for example, again, prototyping can help validate its content and format.
- If the requirements for the report have been described in vague terms, for whatever reasons, prototyping will help define levels of accuracy, optimum format, etc., as the User sees the possible versions of the report and tries to use it.

Setting up the team

The team to carry out the prototyping must be defined well before the activity is due to start, so that the management structures can be put into place in good time. The team should comprise a team leader and two other analysts, who between them will serve the roles of implementing the prototyped model and demonstrating it to the User. There is no absolute need for two analysts, of course, one would be sufficient if the project were not too large; a second team member, however, does provide an objective view of a prototype designed by someone else. The effect of this would be that during demonstration, the analyst would be more sensitive to the User's requirements, and less defensive about the product.

The team leader's responsibility, apart from standard supervisory duties, should be the following:

- Approving the choice of dialogues and reports to be prototyped.
- Agreeing feedback from the demonstration sessions.
- Deciding when to close a prototyping cycle for a product.
- Notifying changes to SSADM documentation, as a result of a prototyping cycle, to the relevant authorities.

- Making the final report to management, summarizing the outcome of the proto-typing cycles and reasons for decisions made.

14.4   Procedures

This section describes the procedures followed in Step 350, and the products of the prototyping exercise.

Define the scope of prototyping

Management will normally have specified the areas, specific dialogues, and output reports that are to be prototyped. This document is not the final word, however. The team's task is to define the scope of the activity, using the outputs from Step 330:

- I/O Structure for each function
- User Role/Function Matrix, defining critical dialogues

The critical dialogues should be studied to see if they are appropriate for prototyping. Confirm the decisions with the Users, and ask for further dialogues that they want to see prototyped. The only constraint on the team's agreement should be budget and timescales.

Some reports also should be prototyped, particularly those using formats dictated by external bodies, such as the Inland Revenue for their PAYE forms, or the bank automatic clearing system (BACS) for their standard forms.

Produce the initial prototypes

The tool for this activity is the Prototype Pathway. This is drawn up for each User Role, and takes that role through the pathway for a particular dialogue, from initial menu, or command, to completion of the function task. Its main purpose is to show the linking of all screens and reports in the prototype.

Figure 14.2 demonstrates a Prototype Pathway for the dialogue Cancel Delegate Booking. This example shows the path the User will take, and identifies each component as a menu, screen, or report. Each component has its own unique identifier and function description.

The components must be identified first, from the logical grouping of dialogue elements (LGDE) constructed during Dialogue Definition, and from the Requirements Catalogue. Once the screens and reports are identified, the pathway is completed by putting them in a logical sequence, and linking them with a Menu or Command Structure.

Once the Prototype Pathways have been identified, they can be implemented on the prototyping tool. The menus may have been implemented already by the team in anticipation of this step, or may be in place in template form, as a feature of the particular tool used.

Once we reach this point, we must begin to think in terms of screen design and ergonomics. Screen design itself is a discipline outside SSADM, and is therefore beyond the scope of this book. Local Installation Style Guides, which should be produced to support this activity, will give guidance to the designer, ensuring that the screens and pathways in these prototypes adhere to local standards.

The designer's responsibility here is to enable the User to concentrate on the content of the screen, rather than be distracted by layout.

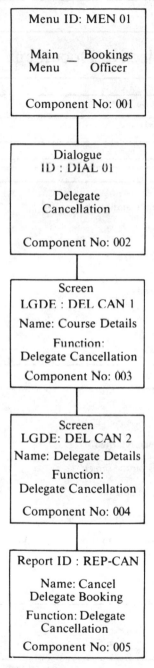

Menu ID: MEN 01

Main _ Bookings
Menu   Officer

Component No: 001

Dialogue
ID : DIAL 01

Delegate
Cancellation

Component No: 002

Screen
LGDE : DEL CAN 1
Name: Course Details
Function:
Delegate Cancellation
Component No: 003

Screen
LGDE: DEL CAN 2
Name: Delegate Details
Function:
Delegate Cancellation
Component No: 004

Report ID : REP-CAN
Name: Cancel
Delegate Booking
Function: Delegate
Cancellation
Component No: 005

Figure 14.2

Prototype Demonstration Objective Document

Document No: 021	Prototype Pathway No: 0001
Function name: Booking Cancellation	User Role: Bookings

Agenda

1. User has not been involved in prototype demonstrations before, so:

(a) Discuss area covered by the prototype.
(b) Explain procedure of prototyping.

2. Arrange for redemonstration.

3. Clarify navigation details of components 3 and 4.

Component No.	Component queries
001	Check OK
002	Is correct screen returned?
003	Are all details on the screen?
003	Is the display intelligible?
004	Are the delegate details all displayed?
005	Does the report need to show new number of places free?

Figure 14.3

During the prototyping sessions, the output data items should be validated, usually against the LDM and data item descriptions. Some data items will be derived from, for example, calculations: the formulae should be recorded with the prototyping documentation, and also on Elementary Process Descriptions and/or Function Definitions.

Prepare for the prototype demonstration

A Prototype Demonstration Objective form should be completed for each pathway identified. Figure 14.3 gives an example for a Prototype Demonstration Objective, for the pathway shown in Fig. 14.2, Cancel Delegate Booking.

The purpose of the document is to force designer and User to understand the objectives of each prototype before the demonstration, rather than play about with the prototype because objectives have not been defined. This will lead to a more fruitful demonstration session.

For each component on the pathway, list its assumptions, purpose, and queries. Identify all the points you want to check with the User, and discuss during the session.

Test data and inputs for each of the prototype dialogues must be prepared to show that the inputs and outputs are as required, and that the processing is accurate. False data must also be prepared to show how the dialogues can handle erroneous input.

The prototypes should be tested against the User requirements, to make sure that the functional scope, inputs, and outputs are all accurate.

It must be remembered during this exercise that the prototype is not the front end of an actual system, but a simulation of the interface only.

Demonstrate and review prototypes

The User Roles involved in each dialogue will have been identified during the early preparatory work on Dialogue Design. Representatives from the relevant User Role should now attend the prototype demonstration of their pathway, with preferably two members of the prototyping team.

The Prototype Demonstration Objective form will be used by designer and User as a checklist of all discussion points for each component.

A Prototype Result Log will be created to record the results of each demonstration. The log will record each user request, for each screen, report or menu, for each version of the prototype.

The results of each component demonstration will be recorded during the demonstration session, and a note made of the kind of change needed to satisfy the requirements.

Every result on the log will be annotated with a code that classifies the change required. There are seven codes that can be applied here:

N No change required.
C Cosmetic changes only. These affect the presentation, not content of the components. Time should not be spent making cosmetic changes only: wait until more substantial changes are to be made, and incorporate the cosmetic alterations then.
D Changes that affect the dialogue only.
P Changes that affect the Prototype Pathway. It may be that the changes will affect other documents, such as I/O Structures for the relevant function.

Prototype Result Log

Prototype Result Log No. 021		Prototype Pathway No. 001	
Function name: Booking Cancellation		User Role: Bookings	

Component No.	Result No.	Result Description	Change Grade
001	01	Satisfactory	N
002	01	Satisfactory	N
003	01	Course date should be shown	D
003	02	Screen cluttered in the centre	C
004	01	Too many delegates shown	A

Figure 14.4

S   The results highlight a weakness in the standards: perhaps they should be investigated with a view to having them altered.

A   Results have indicated that the analysis may be faulty. This is a serious error that will cause a suspension of prototyping activities while the matter is referred to management. At worst it may cause a second pass through some or all analysis stages.

G   A change is indicated which has ramifications beyond the application. It is possible that they would affect the organization's working practices. If that is the case, then again the matter must be referred to management before action is taken.

Figure 14.4 shows the Prototype Result Log for the prototype demonstration of Cancel Delegate Booking.

On completion of the logs, the team leader must make a decision on further action. There are three questions on which to base the decision:

1.   Will further demonstrations be needed, or will no purpose be served by continuing?
2.   Should the timescales be extended, or more resources allocated? If the answer to these is yes, Project Management must be approached to sanction the decision.
3.   Have any problems surfaced that must be drawn to management's attention?

If the prototyping is to continue, the team leader will instruct the demonstrators to make the necessary changes and arrange the new demonstrations.

Perform updates to supporting SSADM documentation

Any changes to SSADM documentation that result from the prototyping exercises should be implemented by the team leader. Such changes may be identified on each pass at a prototype. The products to change may be I/O Structures, Required LDM, Requirements Catalogue, DFM, Enquiry Access Paths, Effect Correspondence Diagrams, or even User Role/Function Matrix. Suggested changes to such documents should be reported and investigated promptly, to see whether they will in fact give the anticipated benefit, or are even practicable. The changes should, when implemented, be tested against the prototype again to see whether they are indeed what the User intended.

When a decision has been made to implement such changes, the team leader must pass details back to the relevant analysts to investigate the ramifications of such changes.

The analysts, having carried out their investigation, will pass details to the Stage 3 technical manager (assuming the appropriate Project Management structure) for approval of the changes, or otherwise.

One aim of Specification Prototyping is to identify new requirements, as well as to validate current ones. If this happens, the team leader can update the Requirements Catalogue directly at the conclusion of the prototyping activity.

Possibly the prototyping activities in Step 350 will result in more functions being identified, for example. If this happens these should be given a Function Definition of their own.

Confirm the specification content

When all the prototyping has been completed, all the changes recorded and implemented, and the Requirements Catalogue amended, the team leader must prepare the report to management on the exercise. This report will cover several related issues:

- Have all the nominated dialogues and reports been prototyped?
- Have all the prototyping objectives been fulfilled? If not, why not?
- How has the Requirements Specification changed as a result of the prototyping exercise? This may be through new requirements, a clarification of identified requirements, or an acknowledgement that earlier requirements were not needed after all. Which SSADM products have been changed—and how—as a result of the exercise?
- What lessons have been learned about prototyping from the exercise? Was the exercise worth while? Should it be approached differently next time?

This report will be the major output from Step 350, along with any amended SSADM products.

Summary

Specification Prototyping is not a method for building a system; rather it is a way of validating the agreed requirements by showing the User an animated model of them.

This has a number of advantages: it enables the User to get an idea of what the system will look like; it acts as a generator of ideas for new requirements; it identifies faulty analysis in time to be corrected before design is started; and most importantly, perhaps, its reliance on the User as a participant ensures User commitment to the new system.

Specification Prototyping takes place in Step 350, following Function Definition and Dialogue Definition. Its output is a report to management and amended Requirements Specification and, sometimes, amendments to other SSADM documentation.

15. Requirements Definition

15.1 Aims of chapter

In this chapter you will learn:
- How SSADM describes the requirements for the new system
- How and when SSADM gathers information about functional requirements and non-functional requirements

15.2 Where Requirements Definition is used in SSADM

The main tool of Requirements Definition, the Requirements Catalogue, is created and amended in the following steps:

Step 110—Establish analysis framework The Requirements Catalogue for the Full Study is created in this step. If a Feasibility Study was carried out, then a catalogue for that study will be carried forward.

The first requirements entered are those identified on the PID, or equivalent document.

Step 120—Investigate and define requirements This is the most significant early step for the Requirements Catalogue. If there is a current system, the step is undertaken in parallel with Steps 130 and 140, in which the current environment is studied; if this is a greenfield site, this step is performed alone. Whichever circumstance applies, the products from Step 120 will drive the Business System Options in Stage 2.

The entries from this step combine fresh requirements from the new system with identified problems in the operations of the current one.

At this point, the emphasis is on functional requirements, but a record may be made of certain non-functional requirements, such as security requirements, major constraints, etc., to aid in the creation and selection of Business System Options.

Step 150—Derive logical view of current services When the current view of processing is logicalized, physical constraints that contribute to identified problems may be resolved. If that happens, then record the facts in the Requirements Catalogue.

Step 210—Define Business System Options The Business System Options are identified from the entries in the Requirements Catalogue. Suggested solutions to the problems and requirements that feature in the BSOs should be recorded.

Step 310—Define Required System Processing } The selected BSO may not address
Step 320—Develop Required Data Model

all of the requirements previously identified. The catalogue should be annotated to explain which requirements have not been met, and why. It is possible that at a later stage, e.g., Technical System Options, they may be addressed after all.

Expand the Requirements Catalogue in detail during these two steps, entering details of new requirements that may surface. Cross-refer to DFD and LDS elements that support particular requirements.

Non-functional requirements, such as retention requirements, access, security requirements, etc., are described in Step 320.

Step 330—Derive system functions   Update and enquiry requirements are annotated in the Function Definition in this step. Cross-refer Requirements Catalogue entries to the Function Definitions. There will not necessarily be a one-to-one correspondence between requirements and functions.

Step 350—Develop Specification Prototypes   In this step we sit down with the Users and make use of the prototyping techniques to clarify—and improve—their understanding and ours of the requirements.

If further requirements are identified during the process, these are entered into the catalogue.

Step 370—Confirm system objectives   The Requirements Catalogue entries are thoroughly reviewed, along with the Required LDM and Function Definitions, for completeness and accuracy. In this exercise we are making sure that all requirements, especially the non-functional requirements, have been identified and described fully.

Step 410—Define Technical System Options   When we reach this step, most of the requirements to be met by the new system will have been identified and addressed. However, there will still be some that have not been met by the logical model, but are more properly in the domain of Technical System Options. Service levels and technical requirements across the business system may come into this category.

Step 510—Define User Dialogues   Any requirements to do with dialogues are resolved in this step. See Chapter 13 for a fuller explanation of the related requirements.

Use in SSADM

Requirements Definition is one of the driving activities of SSADM. The emphasis of the method is on the future system, whether it is one to replace a current system, or whether it is a greenfield site.

There is no one rigid Requirements Definition technique to describe; rather it is a continuous process that involves constant User consultation, carried out from the beginning of the project up to the end of Logical System Specification.

15.3   Relationship with other techniques

As stated in the introduction to this chapter, Requirements Definition is not a discrete technique like, for example, Data Flow Modelling. Rather, it makes use of a variety of skills to identify the requirements, record them in the Requirements Catalogue, and document their solution.

Any SSADM technique can use the Requirements Catalogue entries, or cause amendments to be made; it is the project repository of requirements information. As well as the core SSADM techniques, Requirements Definition also has strong links with the non-SSADM techniques outlined in Project Procedures (see Chapter 2).

The principal procedure techniques related to Requirements Definition are:

- *Capacity Planning*   This is required to ensure that there is sufficient capacity to meet the application's requirements. Also, it is necessary to ensure that meeting the new application's requirements will not seriously degrade current services.

  Using Capacity Planning techniques we can paper test the required service levels against suggested technical environment descriptions.

- *Risk Analysis and Management*   This technique is intended to identify and guard against likely security threats to the information system, whether from terrorist threat to the premises, from 'hackers' illegally accessing the data, from fire threat or from data corruption.

  Requirements Analysis should identify those areas most likely to be vulnerable, and interact with the Risk Analysis techniques to see best how to meet them. There is not a fixed step in SSADM where security and control considerations must be addressed, but the questions must be posed in parallel with the SSADM project.

- *Testing*   Although physical testing of the system occurs after SSADM's involvement with the project has finished, the Requirements Catalogue makes testing objectives easier. Every requirement, whether functional or non-functional, must have a quality criterion which is measurable and quantifiable. These criteria must be entered in the catalogue, and form the basis for subsequent test design.

- *Training and documentation*   The analyst must be aware of the need for User skills to be developed if the system is to work, however technically excellent it may be. The two tools to develop User understanding are training and clear documentation. Requirements of Users and support staff must be noted and addressed.

15.4   Requirements Definition

This section describes the activities involved in Requirements Definition, including the Requirements Catalogue, and the distinctions between functional and the different kinds of non-functional requirements.

Identifying requirements

Another technique applied to requirements is that most basic skill of the systems analyst: fact-finding. This book is not an appropriate place to describe the various fact-finding techniques, but will point out the features that the analyst is aiming to elicit with reference to requirements.

When carrying out the fact-finding exercise, the analyst must highlight the following points:

- What is required from the new system?
- Which User will 'own' this requirement?

- Why is it required? Is it vital, useful, or a nice idea? Some projects will fail if a particular requirement is not implemented, and be unaffected by others.
- What measures can be applied to the requirement? Do not accept a requirement that cannot be quantified by some appropriate measure. This measure may not be forthcoming at first identification; if that is the case, record the requirements in the Requirements Catalogue and return to the question of measurement later on.

Requirements Catalogue

This document is the repository of all requirements information. The first entries are made during Feasibility, and carried forward to the start of the Full Study.

All entries are regarded as provisional, and should be returned to, and revised, as often as need arises up to the end of Logical System Specification.

The emphasis should always be on the future system, rather than on the current one. If there is no current system, the first requirements entered here will form the basis of Logical DFDs, LDM and BSOs, as envisaged by the Users of the new system.

Figure 15.1 illustrates an entry for the Requirements Catalogue for SS plc.

Functional requirements

These are the requirements that, from the User's point of view, perform the activities that run the business. These include all updates to master files, enquiries against the data on file, producing reports, and communicating with other systems relevant to the business activities.

Non-functional requirements

These requirements define the performance levels of the business functions. These include such features as response times for on-line transactions, turn-round time for batch inputs, and levels of accuracy. Other non-functional requirements concern such issues as security, recovery and back-up in case of breakdown. Some of these, such as back-up facilities, may be common across the whole business system, while others, such as response time, may be specific to a given application. Whether the requirement is local or global should be noted in the Requirements Catalogue entry.

The non-functional requirements are likely to be reviewed and revised several times during the study as the analyst learns more and more about the system. Step 350, Develop Specification Prototyping, is a likely place for amendments, as the User sees precisely what is offered, and understands what is possible.

Quantifying requirements

To avoid ambiguity, and to give a basis for testing, requirements should be given some form of quality measure. The act of quantifying the requirement helps to focus the analyst's attention on the specific requirement. Vagueness suggests that the requirement may not be fully understood.

This is obviously easier in the question of non-functional requirements. Some functional requirements may not lend themselves to quantification easily. For instance, if a requirement is to improve customer satisfaction, how do we measure something as intangible as that? We must settle on some criteria that tell us what has been achieved. A possible measure is that we receive a certain proportion of verbal or written compliments, or that our business with present customers increases by a set percentage.

Requirements Catalogue Entry

Source	Owner	Requirement ID	Priority H
Booking Officer	Course Mgr.	9	

Functional requirements

Provide Booking Clerks with on-line access to names of delegates on Waiting List for a given course, to speed up replacement of place after delegate cancellation.

Non-functional requirement(s)

Description	Target value	Acceptable range	Comments
Response Time	3 seconds	3 - 6 seconds	
Service Hours	9.30 - 5.00 Mon - Fri		
Availability	90%	85 - 90%	

Benefits

Will speed up process of finding names and creating an offer. Will free Booking Clerks for other pressing work.

Comments/suggested solutions

To fit terminals on to two clerks' tables, with on-line access to Waiting List names, in first-come, first-served order, or in Branch order.

Related documents

Required System DFD, Process Box 5

Related requirements

7. To offer places to Delegates from same Branch as Delegate who cancelled.
11. To identify cancellation penalties

Resolution

Figure 15.1

Specifying requirements

The entries in the Requirements Catalogue express the User's perception of what the new system is to achieve. They are not precise enough, though, to act as a specification for the new system.

Each entry needs to be supported by the more rigorous expression of the SSADM techniques of Entity/Event Analysis, Logical Data Modelling, and Function Definition. These will all be based on the entries in the Requirements Catalogue, and can be traced back to the appropriate entry, but are more precise than the textual descriptions in the entries.

Entries which are not met by these techniques should be carried forward into specification. They may be addressed by other techniques, or by the particular technical implementation.

Summary

Requirements Definition is a continuous process, not a simple step activity. It is the driving force in a SSADM project, always forcing the emphasis on to the new system, rather than the current environment.

The principal tool in Requirements Definition is the Requirements Catalogue. This is started at Project Initiation, and still updated even during Physical Design.

From the first steps, we are identifying the functional requirements and, to an extent, some non-functional requirements. The latter are examined in more detail as the study proceeds.

All requirements should be quantified to provide a basis for planning and evaluating tests later in the cycle.

Participants in Requirements Definition are the analyst, designer, User and also the service providers, who define the non-SSADM requirement techniques such as Capacity Planning and Risk Assessment.

16. User options

16.1 Aims of chapter

In this chapter you will learn about the three key places in SSADM where the User is presented with a selection of options, and must choose which way the project will progress. You will learn about the considerations that influence the User's choice, and how the selection is made.

16.2 Where User options appear in SSADM

The options are of two basic kinds: Business System Options, which provide us with the scope and objectives of the project, and Technical System Options, which define the technical environment in terms of hardware, software, and development approach.

In each case, the analyst prepares a set of options, with full supporting documentation, and Cost/Benefit Analysis. The User, after a presentation of each option, considers their relative merits and makes a selection. This choice may be to abort the project at that point, or to proceed along a certain path. User options are defined or selected in the following steps:

Step 030—Identify Feasibility Options
Step 210—Define Business System Options
Step 220—Select Business System Option
Step 410—Define Technical System Options
Step 420—Select Technical System Option

16.3 Feasibility Module

The Feasibility Study is usually instigated as a part of a Strategic Study, carried out some time earlier. This Strategic Study will have identified areas for computerization and redesign, and accorded priorities to the various project areas.

The objectives of Feasibility are to investigate the project area and identify (a) whether or not the project is technically feasible, and (b) whether a sound business case can be made for pursuing it.

The end of Feasibility is a set of options presented to the project board, each making a different recommendation about the path to be followed.

It is possible that if circumstances have changed since the Strategic Study, that no good business case can be made for pursuing it. If so, the recommendation will be to abort, or at least postpone, the project there.

Each option meets the requirements defined in the Project Initiation Document and identified in the earlier parts of the study, at least to a specified minimum level.

The analyst will also produce for each option outline project plans as to how to proceed.

Business System Options and Technical System Options are described in detail below, so I shall not dwell on them here, other than to say that, at Feasibility, each option will contain elements of Business System Options and Technical System Options.

Feasibility Options are high-level only, and are far less detailed than the full Business and Technical System Options. The outcome of Feasibility will be a recommendation to proceed, along the lines suggested, but a Full Study will follow, and that is where the detailed analysis of the costs, benefits and impacts of the options will take place.

16.4   Business System Options—Requirements Analysis Module

Stage 2, in Requirements Analysis, is concerned with the selection of the Business System Options (BSOs). The two steps cover all BSO activities, and form the major input to the next Module, Requirements Specification.

A BSO describes a suggested new system in terms of its functionality and its boundary. Inputs, outputs, processes, and data are described, just as in the Current Environment Description. Its aim is to help the Users choose, from all the listed requirements, just what they want their new system to do.

Its input are the Requirements Catalogue and Current Services Description. Development of the BSOs is done in consultation with the Users. When the Users are presented with the options there should be no surprises.

The format of the BSOs is a text description of the boundary and functions to be performed, although these will probably be illustrated with high-level DFDs and an LDM. At option development these techniques are not used in detail, but rather give a broad-brush view of the system. They are considerably enhanced when the selected option is built up into the specification of requirements.

Technique

First, draw up a list of about six BSOs, covering a range of the requirements identified in Stage 1. The range should cover:
- One option that covers the stated minimum requirements, and no others.
- One option that covers every new requirement.
- Up to four options that each cover the stated minimum requirements and a different set of the other requirements.

The six options will then cover six different boundaries and six different functionalities, yet all will cover the essential requirements identified as necessary for the new system.

These first options should be skeleton only, stating no more than what the option is going to cover. They should then be expanded to describe the impact upon the business and the non-functional requirements that have been met. The impact upon the business should be expressed in terms of the priority of the BSO in relation to the IS strategy.

As the BSO is expanded further, bring in details of expected volumes: volumes of data; volatility of data; and frequencies of key tasks, especially at peak times.

Also describe the BSO in terms of project development, looking at the following aspects:

- Cost/benefit of proposed option
- Impact Analysis of implementing the BSO
- Timescales for development and construction

Finally, when the six skeletons are produced, the analysts and Users together must begin eliminating obvious non-starters. The criterion may be cost, organizational impact, or inadequate functionality. Whatever reason, the list should be reduced to two or three options for the presentation.

Having whittled down the list, we now expand the remaining options with greater detail. The prose description of the options will be longer and more explicit, defining more closely the functions performed, inputs/outputs, data, and processes. The selection of processes for batch and on-line processing will also be defined, with a (rough) idea of service levels that can be provided.

The selection considerations are defined in greater detail. These take into account:

- Costs/benefits
- Constraints
- Impacts on existing systems
- Plans/timescales for subsequent SSADM activities, and implementation of the system
- Organizational impacts and implications

The differences between the individual options are unlikely to be radical. They may depend on a trade-off between cost, security, functionality, and service levels, with seemingly marginal differences between any two. This is obviously acceptable, as long as the differences are not so marginal as to be not real.

Selection of BSO

In Step 210 we have, with the Users' help, defined the options from which they will choose. In Step 220 we present the options to the body which will decide on the future course of the project.

The presentations may be made to the Project Board directly, or to a User body that has been empowered by the Board to make the decision. Standard presentation techniques, not specific to SSADM, are employed by the analyst for the exercise. The four basic activities are:

1. Prepare the presentation.
2. Deliver the presentation.
3. Help the User decide, and afford clarification as needed.
4. Record the selections made.

At the presentation, the analyst will identify the points of difference between the short-listed options. The review body will probably want to question the team further about distinctions in functionality, service and response levels, costs, and development times. With the answers to these points the Users may then take the decision.

The decision may be to accept one of the options as it stands, or to make a hybrid selection, combining features of two or more options. If this happens, the analysis team must carry out a fresh exercise of costing, justifying, and scoping the new option. This will be done in greater detail than the others, as it will form the basis for the specification for the new system design.

Another decision may be to stop the project at that point, or to reject all of the options and ask for more to be prepared. If that happens, it suggests that the analysts did not carry out the preparation thoroughly enough; by the time that presentations are made, the Users should be familiar with the likely courses to follow and should have made it clear if none were acceptable. However, there may be other sound business reasons for rejecting at that point. If it happens, the analysts must go back on their tracks and examine the Requirements Catalogue again.

Whatever decision is taken, it must be recorded in the project documentation. Details to include are:

- Option chosen
- Reasons for adoption
- Options rejected
- Reasons for rejection of each

These details, plus a detailed description of the selected BSO will be passed, via the information highway, to Stage 3.

Application of BSO

As with Technical System Options what I have described is a set of guidelines to help the analysis team to reach a solution to the requirements. They are not intended to be fixed rules, or an algorithmic technique. Many SSADM techniques are 'soft' rather than 'hard', and BSOs, being creative and forward-looking, are necessarily very 'soft'. If this means that your organizational standards dictate a different approach, then follow that.

What SSADM does stress is the role of the User in choosing, with the analyst's careful advice, the business and functional path that the project will follow.

16.5   Technical System Options—Logical System Specification Module

Stage 4 is concerned with the production and selection of Technical System Options. TSOs are the decision point after BSOs, where the Users are able to decide the future course of the development, even to the point of abandoning the project.

The TSO will define the implementation environment and strategy for the selected BSO. The issues addressed are:

- Specification and description of the hardware, software, and data environment. This will be in generic terms, and will not lead to a tie in to any particular vendor: tenders will not be invited at this stage.
- Confirmation of the functions in the application area, and mode of processing for each.
- Description of the organizational impact and effect on work methods.
- Description of the impact on the rest of the development organization, and the remainder of the project.

Many of these issues may have been addressed already by the Feasibility Study, and/or the IS strategy for the given functional area. If this is so, then the range of TSOs will be constrained by these earlier decisions, which are to be found recorded in the Requirements Catalogue. If there has been no Feasibility Study, and the IS strategy has not tied the system into a particular technical environment, these issues will all be considered.

Inputs to the TSO will be a mix of SSADM and non-SSADM documents. SSADM inputs are as follows:

- Requirements Specification
- Selected BSO (with reasons for selection)

Extra-SSADM inputs are as follows:

- Project Initiation Document (PID)
- Description of current IS environment
- IS strategy
- Standards
- Security standards
- Installation Style Guide

While the SSADM inputs are the drivers of the TSO, the other management inputs will be a source of influence and guidance.

The options themselves will take the form of:

- Outline Technical Environment Description (TED)
- System Description
- Outline Development Plan
- Cost/Benefit Analysis
- Impact Analysis

Each of these will be completed in outline only: with several options to complete, only one of which will go forward, too much effort is required to give full and detailed descriptions for each.

Technique

The procedure for producing and selecting TSOs is very similar to that for BSOs. The difference lies in the objectives of the options, and the constraints that apply. This section describes the techniques briefly, and then looks at the particular constraints that affect Technical (as opposed to Business) System Options.

First, draw up an initial list of six or so options. These may come from methodical discussion of different approaches, or may be generated by a brainstorm. However they are raised, once the six are created in skeleton form, they must be expanded. This expansion though, often requires investigation of suppliers to obtain details of such things as costs, facilities, performance, support, etc. Note: this is not to pre-empt choice of vendors, but to obtain 'ballpark' figures and estimates to present to the Project Board for each option.

If the current system is a manual one, then one of the initial options must be a no-computer option. Similarly, if the current system is a computer-based system, one option to be considered should be a no-change option, i.e. end the project here. At this stage, this choice is unlikely.

Secondly, six TSOs are too many to specify in sufficient detail for presentation. We must, therefore, cut them down to a more manageable number. Three is an acceptable number, but you may find that you want or need to include a fourth viable option. Often this can be done automatically, as some options are obviously more appropriate than others. Always consult with the User, on an informal basis, as to which should be taken forward. It may be that the User asks for features from two separate outlines to be combined into the new option. Any feature which is not acceptable or appealing to the User can be identified and discarded at this point.

Thirdly, once the number of TSOs has been agreed, the skeletons must be fleshed out with detail. The proposed configuration of each will be the central document.

Aspects such as required data volumes are needed in order to assess the viability of a suggested hardware/software configuration, so a rough sizing exercise for each option must be carried out. A non-SSADM technique, Capacity Planning, will help in this task. Capacity Planning is one of the Project Procedures described in Chapter 2.

As the options are developed, changes may be needed to earlier documents, or may be made to the options in the light of earlier work in the project. Some form of change control mechanism should be set in place, so that changes can be implemented and recorded without inconsistencies and inaccuracies creeping in.

Changes to earlier documents might take the form of amending ELHs to assess or alter supervisory events and constraints, or changing/creating entities on the LDS, to take account of the proposed configuration's capabilities/constraints. Other products may be susceptible to change during the TSO development. The information highway must be able to pass the changes to the relevant control authorities to preserve the system's integrity. It may even be that as the selected TSO is being refined and expanded later, a complete second pass through Stage 3 is required to ensure that the design process has full information to work on.

Fourthly, the options to be presented will be prepared according to installation standards, but must contain certain information to allow the Project Board/User body to make a sound decision. The features that must be included are as follows:

OUTLINE TECHNICAL ENVIRONMENT DESCRIPTION (TED)

This will describe the hardware, software, development environment, system size, and fallback/recovery proposals. The TED explains to the User how the system works, and how it will be developed. It should contain enough information to allow reasonable costings to be made. At this point, however, the TED should be in outline only. After selection, the chosen TED will go forward as the major document.

SYSTEM DESCRIPTION

This may have been covered already in the BSO. Its province is the solution to items in the Requirements Catalogue. To reduce effort in the preparation of the TSO, you may wish to use SSADM products such as the Function Definitions, annotated Requirements Catalogue, and Required DFM and LDM.

IMPACT ANALYSIS

This is concerned with the impact of the proposed system on the User's environment. The issues involved are general systems analysis and project management issues, not particular to SSADM. The concern is not with the technical feasibility of the technical system, but with the organizational implications. The issues involved cover the following topics:

- Organization and staffing
- Changes in operating procedures
- Savings from replaced equipment, or from maintenance no longer required
- Implementation considerations (from staffing point of view)

- Training requirements
- User manual requirements
- Testing requirements
- Take-on requirements
- Advantages/disadvantages against the other TSOs

OUTLINE DEVELOPMENT PLAN

This plan provides management with proposed strategies for developing the particular options, with ideas of timescales and resources required. From this information, management can prepare plans and compare the development overheads of the respective options. The plan should contain information under the following headings:

- System design
- Program design/coding/testing
- Procurement (if relevant)
- Implementation

The precise contents of these sections will depend upon the circumstances and standards prevailing at the installation. For each, though, some statement should be made that will enable management to carry forward the planning and costing of development into the next stage.

For the estimating exercises—both time and cost—the Estimating Subject Guide referred to in Chapter 2 will be necessary, unless local estimating methods are prescribed.

COST/BENEFIT ANALYSIS

This provides the Project Board with quantifiable selection criteria. It deals with the financial specification of each option, and so is the part that often decides the board's choice.

It covers a number of areas, under both costs and benefits, the principal ones being:

- *Development costs, or non-recurring costs*   These can be calculated from the TED, after consultation with a number of vendors. The outline development plan provides costs of resources needed for the development.
- *Operating costs, or recurring costs*   The TED and Impact Analysis provide data for calculating these.
- *Tangible benefits*   These are measurable financial benefits of implementing the new system. Examples are money saved in better stock control, or higher interest from improved cash flow, or increased profit margins.
- *Intangible benefits*   These are benefits that are harder to quantify, but do result from implementing the new system. Such things as improved customer service, higher staff morale, and better work environment are included under this heading.

  While a precise figure cannot be put on such benefits, some yardstick for measuring should be attempted, such as adding one per cent business for satisfied customers, or lower staff turnover, saving X per cent recruitment costs per annum for higher morale

Selection of TSOs

The procedure here is the same as that for BSOs. There are four activities involved:
- Prepare the presentation
- Make the presentation
- Assist the Users/Project Board
- Record the selection decision

As with the BSOs, the final selection may prove to be a hybrid of features from two or more options. It is unlikely that a decision will be made immediately the presentation has finished: there will be too many details to consider. However, a date in the near future should be agreed for a decision, so that the development plans, and therefore costs, are not impacted by any delay.

The decisions should be recorded and the TSO and TED updated accordingly. The TSO documentation will be filed away now, and the TED of the selected option expanded to be carried forward to Physical Design.

The selected option will be examined again from the point of view of Capacity Planning to make sure that it will meet the service-level requirements. If they cannot be met, another decision must be made:
- Propose a new architecture with greater capacity.
- Reduce the service-level targets.
- Propose alterations to the Requirements Catalogue.

Whichever body accepted the option must meet to decide upon the choice to make here, and decide what extra work needs to be undertaken.

Summary

There are three places where the Users take a crucial stop–go decision in a SSADM project:
- Feasibility (Stage 0)
- Business System Options (Stage 2)
- Technical System Options (Stage 4)

The Feasibility Options include elements of BSOs and TSOs, and may set the decisions for the subsequent project development.

BSOs give the User a set of options that define the scope and functionality of the new system.

TSOs give the User a set of options that show how the BSOs are to be implemented. This is the option point that provides the planning and costing data for the Project Board. The development of the TSOs may, under some circumstances, cause a second pass at Stage 3, so requiring a configuration control procedure to be set up.

The output from the TSO forms the input to Physical Design at Stage 6.

17. Logical Database Process Design

17.1 Aims of chapter

In this chapter you will learn:
- How to derive and draw Enquiry Process Models
- How to derive and draw Update Process Models
- How to recognize structure clashes
- How to classify success units

17.2 Where Logical Database Process Design is used in SSADM

Logical database process design (LDPD) is performed in the following steps.

Step 520—Define Update Processes   In this step we complete the Entity Descriptions with state indicators. Enquiry Access Paths and Effect Correspondence Diagrams are converted into structure diagrams that reflect the processing logic of given events, using Jackson diagramming techniques.

Step 530—Define Enquiry Processes   The Enquiry Access Paths from LDM are taken as the input structure to the enquiry process, and the I/O Structure from the Function Definition is transformed into an output structure. The two structures are merged to form the enquiry process.

Use in SSADM

LDPD is used to define the processing requirements on the logical data model. It translates the information gathered during Requirements Specification into a processing logic which is implementation independent. It should also define the logic so that it can be easily maintained after implementation.

The Logical Process Specifications are fed into Stage 6, Physical Design. If a 4GL is being used to generate the code, the specification from LDPD should be sufficient input; if a 3GL environment is to be used, Physical Process Design will translate the Logical Specification into a form that can be coded to access the physical database. The expectation of SSADM is that it will be implemented, in a 3GL environment, by means of JSP (SDM in a government installation). This is not to say that every SSADM design must be so implemented, but the CCTA publication on the 3GL interface does bias development strongly in that direction.

The process models consist of:
- A diagram, either Enquiry or Update Process Structure
- An operations list as supporting documentation

Inputs to LDPD are as follows:
- Effect Correspondence Diagrams
- Enquiry Access Paths
- Entity Life Histories
- Function Definitions
- I/O Structures
- Required System Logical Model

Outputs from LDPD are as follows:
- Enquiry Process Models
  —Enquiry Process Structure Diagram
  —Enquiry Process Structure operations list
- Update Process Models
  —Update Process Structure Diagram
  —Update Process Structure operations list

Figure 17.1 illustrates the relationship of LDPD with other SSADM techniques.

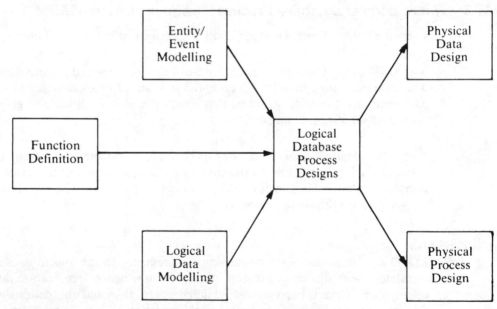

Figure 17.1

17.3   Procedures

LDPD is concerned with those parts of a function that are concerned with the Update or Enquiry Process components. The processing requirements of the business functions are translated into the standard sequence/selection/iteration structure, together with the operations that are carried out to perform the function, or, usually, subfunction.

Enquiry and Update Processes are approached through the Enquiry Access Paths and Effect Correspondence Diagrams. They seem at first sight to be independent, but in fact may prove to be closely connected, in two ways:

1. An Update Process may need to be extended to produce an output report. If so, the Enquiry processing procedures will need to be followed.
2. An Update Process is often introduced by an Enquiry Process. Investigate with the User to see whether or not this situation is regarded by the User as two discrete operations or one. The answer will tell you whether to define one function or two.

Enquiry Processes

The basic description of the procedures to follow is simple: unfortunately, simplicity is not often to be found in the information world, and so the application of these procedures leads to a number of possible complications. I shall begin by describing the simple procedures.

Enquiries come in two basic forms: structured, enquiry functions, and *ad hoc* enquiries, which are more loosely specified. *Ad hoc* enquiries are, by their nature, difficult to model, and often hard to anticipate; I shall therefore only look at the formal Enquiry Functions here.

Specify the enquiry name   Each enquiry must be uniquely identified, the name being used also in Function Definition and Physical Process Specification.

Specify the enquiry trigger   The trigger consists of the data items input, as identified in Logical Data Modelling Enquiry Access Paths. If the enquiry requires a simple navigation around the data model for specific entity occurrences, the trigger will be the key of the entity which is the entry point.

If, on the other hand, the enquiry is to find all occurrences of a class or category, such as all tutors not assigned on a certain date, or all branch managers due to be

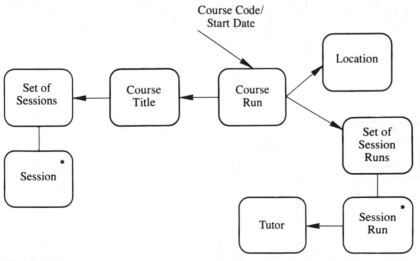

Figure 17.2

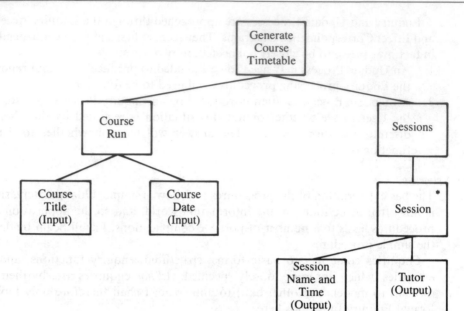

Figure 17.3a

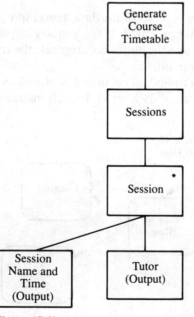

Figure 17.3b

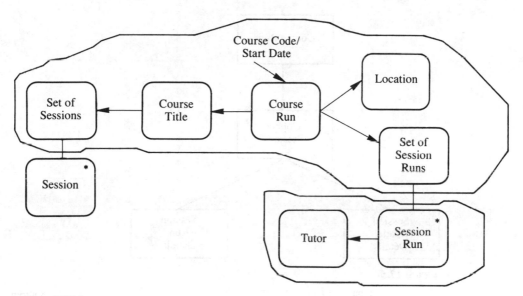

Figure 17.4

invoiced, then the trigger will be the name of the data items and the search criteria, which will be the specified values for those items.

Specify the Enquiry Access Path In Step 360 we created Enquiry Access Paths from the LDS, one for each specified enquiry. Figure 17.2 is the Enquiry Access Path for the requirement, Generate Course Timetable.

Specify the enquiry output In Step 330 we completed I/O Structures for identified requirements. Figure 17.3a gives the I/O Structure for the same requirement, Generate Course Timetable. Strip off the input elements of the structure, so that what is left is the structure of an output report (Fig. 17.3b).

Group accesses on the Enquiry Access Path Group the accesses on the Enquiry Access Path that are in a one-to-one correspondence. Figure 17.4 shows this for our enquiry.

Convert to structure notation Convert the Enquiry Access Path to a structure diagram Jackson-style. In simple cases, this should be an automatic procedure, but more complex EAPs may be more demanding of your judgement. You may need to insert new nodes to conform with the rules of the structure model. Figure 17.5 shows the structure derived from the grouped EAP for our timetable.

Identify correspondences between input and output data structures The EAP is the input structure to this procedure, and the modified I/O Structure is the output structure. They have points of one-to-one correspondence on them. Mark these points with correspondence arrows. The points of correspondence on the two diagrams must occur in the equivalent levels. If the correspondence arrows cross, you

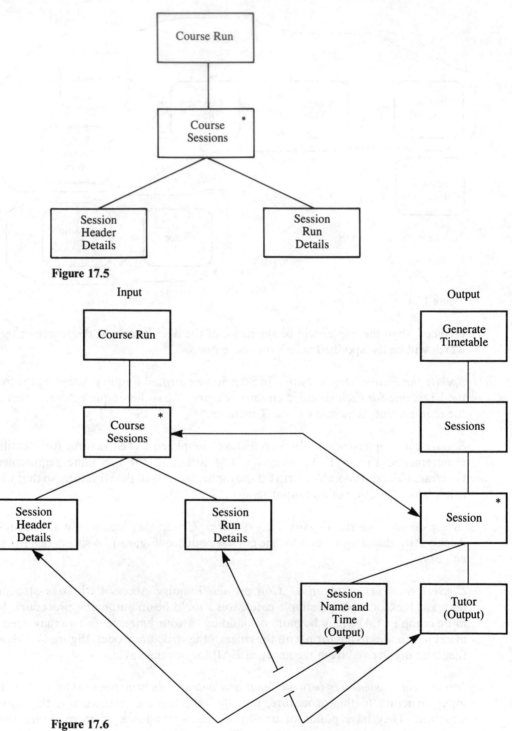

Figure 17.5

Figure 17.6

may have a *structure clash* (see Sec. 17.4). In Fig. 17.6 we see the two structures, and where the correspondences naturally occur.

Merge the input and output data structures   Enquiry Process Modelling consists of taking the input and output structures and merging them, according to the correspondences between them, as shown in Fig. 17.7.

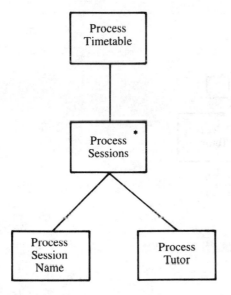

Figure 17.7

List the operations and allocate them to the structure   The operations that can be attached to the structure are, obviously, 'read' operations. Unlike the operations on the ELHs, they can be put under structure boxes as well as elementary boxes. As integrity checks, include operations to abort the process if the state indicators are not in a valid state.

The permitted operations are:
- Read ⟨entity-type⟩ by key: read the database entity using the input key value.
- Define set of ⟨entity-type⟩ matching input data: define a set of ⟨entity-type⟩ entities, the members of which match the criteria in the input data.
- Read next ⟨entity-type⟩ in set: read the next entity of specified type from the currently defined set. There should always be a 'define set' operation before this one is used.
- Read next ⟨detail⟩ of ⟨master⟩ [via ⟨relationship⟩]: read the next detail (specified entity type) of current occurrence of master (specified entity type). The optional 'via ⟨relationship⟩' applies if there is more than one relationship between the two entity types.
- Read ⟨master⟩ of ⟨detail⟩ [via relationship]: read the master (specified entity type) of current occurrence of detail (specified entity type). The optional 'via ⟨relationship⟩' applies if there is more than one relationship between master and detail.

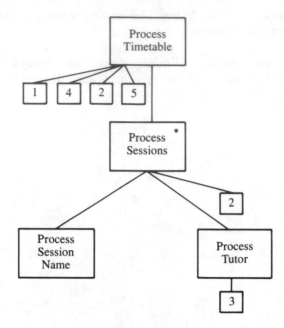

Operations

1. Read Course Run by key
2. Read next detail Session Run of master Course Run
3. Read master Tutor of Session Run
4. Read master Course Title of Course Run
5. Read master Training Centre of Course Run

Figure 17.8

Figure 17.8 shows the operations list and the allocations for the Generate Timetable Enquiry structure.

Specify integrity error conditions   Integrity problems are usually associated more with Update Processes, but enquiries are obviously valid only if the database is correct, so note any errors here.

The commonest kind of integrity failure that can be tested is the absence of a record that should be there. The record may not be in the valid state for the particular enquiry, or indeed, it may not be on the database at all. If that is the case, then there should be a ⟨fail-if...⟩ operation built in to the operations that access the database. These should be shown as operations on the list, and inserted immediately after every database access.

Specify error outputs   With the User's help, design the error report, whether printed or displayed. It is up to the User whether the reports form part of the valid output report, or form a separate report on their own account.

Walk through the structure   Again with the User, conduct a structured walkthrough of the Process Model, to ensure that it meets the requirements and can actually retrieve all the information necessary to answer the enquiry.

Update Processes

Our input to this is the result of Entity/Event Modelling. Every identified event has an Update Process associated with it. Whereas Enquiry Processes are derived from Enquiry Access Paths, Update Processes are derived from Effect Correspondence Diagrams.

The source of the I/O Structures is more complex than for the Enquiry Processes, in that each event may be input via more than one function. Each Function Definition must be checked to ensure that the data items for the event are contained in the input structure.

Specify the event name   The event must be specified by a unique name that is used in Function Definition, Entity Life Histories, Effect Correspondence Diagrams and Physical Process Specification.

Specify event data   The event data means attributes—normally the key—which act as the entry point to the Logical Data Model, and any additional updating information. This additional information may be new values for current attributes, or new key and attribute information for insertion of a new record.

Specify the Effect Correspondence Diagram   This is carried out in Entity/Event Modelling in Stage 3. It encompasses the identification of all the effects for each event. By 'effect', I mean all entities that are affected.

Specify event output   This means all output data items from an event, excluding any error messages. The output may be trivial, such as an acknowledgement that processing has been successfully completed, or it may be in the form of some structured, necessary output, which needs to be modelled on the same lines as enquiry output.

Extend the ECD with enquiry-only entities   If, for an update to be performed, reference has to be made to other entities, include those reference entities on the ECD.

Group effects in one-to-one correspondence   This is described in Chapter 12. Draw a box around each set of entities in one-to-one correspondence, and award each group a meaningful name as a process. Note that there may be one or two entities that are not included in any such grouping. That is acceptable. They too should be given a meaningful process name. Figure 17.9 shows the Effect Correspondence Diagram for the event Remove Delegate.

List operations   For each entity affected, list the operations that apply at that stage

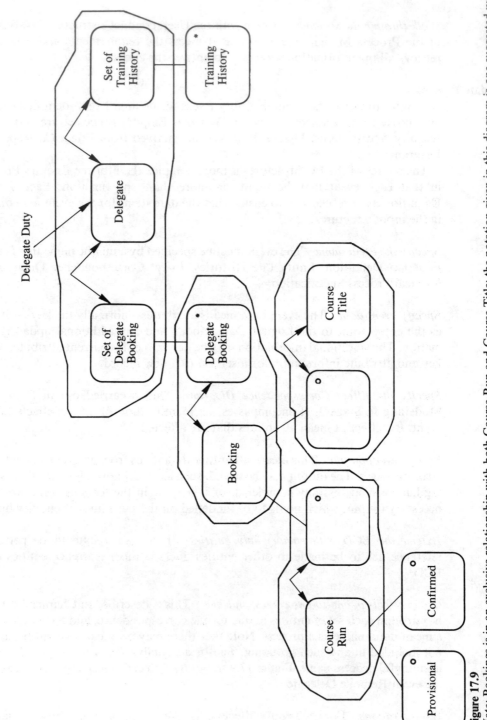

Figure 17.9

Note: Bookings is in an exclusive relationship with both Course Run and Course Title; the notation given in this diagram is one suggested way of modelling this situation in an ECD.

in its life. The classes of operations should, for each entity, be in the following order:

1. A read operation
2. Operations to raise error messages on invalid state indicators. Invalid SIs are identified from the 'valid previous' SIs on each entity. Note that the SIs are added to the ELHs for each entity as the first task in the current step, Step 520.
3. Operations from the relevant effect on the ELH are expanded and added. 'Gain' and 'lose' operations are not carried forward. The ELH operations are expanded as shown below. The additions, which specify the particular occurrence affected, are shown in the curly brackets. Phrases in square brackets are optional.
   (a) Store keys {of ⟨entity⟩}.
   (b) Store ⟨attributes⟩ {of ⟨entity⟩} [using expression].
   (c) Store remaining attributes {of ⟨entity⟩}.
   (d) Replace ⟨attribute⟩ {of ⟨entity⟩} [using expression].
   (e) Tie {⟨entity⟩} to ⟨master entity⟩ [using ⟨relationship⟩].
   (f) Cut {⟨entity⟩} from ⟨master entity⟩ [using ⟨relationship⟩].
4. An operation to reset the entity's state indicator, using the valid 'set to' value.
5. An operation to write the entity.

There are some integrity checks upon operations:

- Every 'read' operation should be followed by an operation to check the SI value, unless all values are valid for that effect.
- Every 'store key' operation should follow a 'create' operation.
- If any attribute or the SI value of an entity is changed, there must follow a 'write' operation.

The operations list for Remove Delegate is as follows:

1. Read ⟨delegate⟩ by key.
2. Fail if SI < 1.
3. Cut ⟨Delegate Booking⟩ from master ⟨Booking⟩.
4. Read master ⟨Booking⟩ of ⟨Delegate Booking⟩.
5. Read master ⟨Course Run⟩ of ⟨Booking⟩.
6. Replace ⟨No-prov⟩ of ⟨Course Run⟩ with ⟨No-prov − 1⟩.
7. Replace ⟨No-conf⟩ of ⟨Course Run⟩ with ⟨No-conf − 1⟩.
8. Write ⟨Booking⟩.
9. Fail if invalid SI.
10. Cut ⟨Delegate Booking⟩ from master ⟨Delegate⟩.
11. Read master ⟨Course Title⟩ of ⟨Booking⟩.
12. Delete ⟨Delegate Booking⟩.
13. Get next ⟨Training History⟩ for master ⟨Delegate⟩.
14. Cut ⟨Training History⟩ from master ⟨Delegate⟩.
15. Delete ⟨Training History⟩.
16. Delete ⟨Delegate⟩.
17. Set SI to null.
18. Read next ⟨Delegate Booking⟩ of master ⟨Delegate⟩.
19. Set SI to valid value.
20. Replace ⟨Nos-booked⟩ of ⟨Course Title⟩ with ⟨Nos-booked − 1⟩.
21. Write ⟨Course Title⟩.
22. Read master ⟨Course Title⟩ of ⟨Booking⟩.

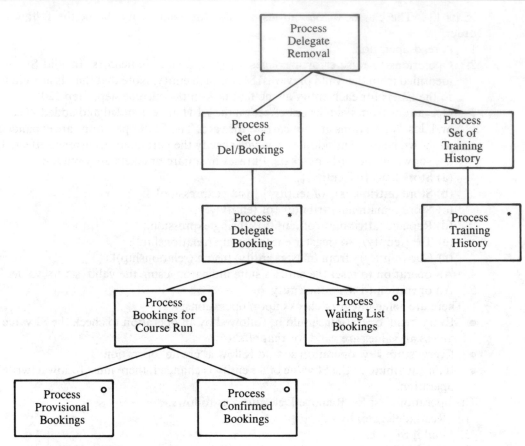

Figure 17.10

Convert to structure notation The ECD, with the effects grouped together, is converted into structure diagram notation. Figure 17.10 shows the ECD for Remove Delegate converted to a structure diagram.

Allocate the operations to the structure Operations should be allocated in the order indicated earlier in 'List operations'.

Allocate conditions to structure Each selection and each iteration on the structure should have a condition governing the path to be followed. The condition may test the value of a state indicator or another value; it may test for the presence or absence of a record. Figure 17.11 shows the operations and conditions added to the Remove Delegate structure.

Specify integrity error conditions Each update process must check that the relevant portion of the database is in a valid state before and after any updates are made. One kind of integrity check is the state indicator check made after any read operation. Other checks may involve examining a value of one attribute, or a combination of attributes, or relative values of two or more attributes. Such checks usually involve

Conditions

C1   While not end of set.

C2   If master $\langle$ Course Run $\rangle$ present.

C3   If Del-book-type = $\langle$ P $\rangle$.

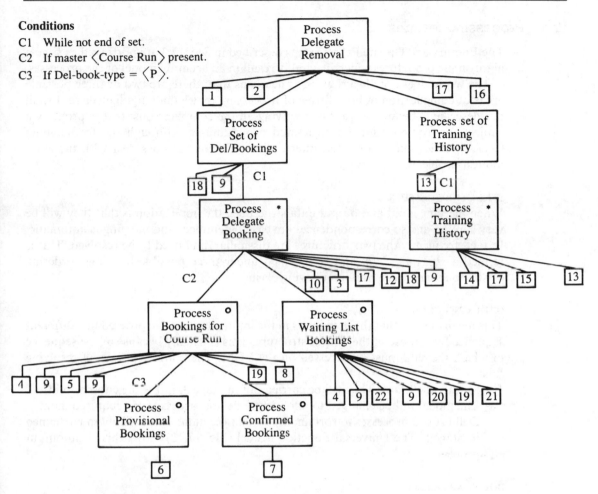

Figure 17.11

applying business rules to the entity, so that some actions may not be taken if certain conditions are not fulfilled.

Each integrity error condition must be recorded on the operations list.

Specify any error outputs   The error outputs may be in the form of coded reports (i.e., giving a code number or letter to be given a meaning from a table), or an integrity error report. They may take the form of a separate report, a message to the screen, or a part of the standard output from the process. Whichever, it is a good idea to begin defining the error output at this point, to prepare the way for Physical Design. It may be useful even to begin the Physical Design of the error output now.

Walk through the structure   This is a quality check on the product, to be held with the User, to ensure that it successfully models the Update Processing requirements. The quality checks on this will mostly centre on the logical sequencing of the processing, and on the correct specification of integrity errors.

17.4   Processing features

The Enquiry and Update Process steps described in Sec. 17.3 are portrayed as simple algorithmic procedures which inevitably result in an accurate model of the processes.

The world does not always work as neatly as that; there are two or three possible areas of complication in both forms of processing that I shall highlight here. I shall not describe in detail the ways of resolving all the complications: that is properly a subject for a publication on structured programming, although the forthcoming CCTA publication on the 3GL interface (see Chapter 2) does deal with the commonest problems.

STRUCTURE CLASHES

When merging input and output data structures, the implication is that they will be very similar, and so correspondences are easily identified, and merging is automatic. In fact, frequently the two structures are dissimilar, and need to be resolved. This is termed a *structure clash*, and is classified under three possible headings: ordering clash, boundary clash, and interleaving clash.

ORDERING CLASH

This means simply that the data items in the input structure are ordered in a different sequence from those in the output structure. This is usually because of the sequence in which the data must be accessed on the LDM. There are two ways of resolving this:

1. Modify the LDM. This will be carried out in Stage 3, if it is the chosen solution. The model will be changed to allow more flexible access to the required data.
2. Define extra processes to sort and/or format the data. This is likely to be defined in Stage 5. The Universal Function Model (Fig. 17.12) shows this as an output process.

BOUNDARY CLASH

Here, the data elements are grouped differently in the input structure and the output structure. This is because on the LDM the detail entities are grouped beneath the

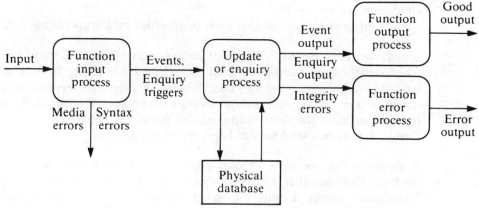

Figure 17.12

appropriate master, while on the output structure the grouping is done on ergonomic principles, relating to the output display.

There are, again, two possible solutions to this problem:

1. If you have an automated tool, such as an application generator or a report generator, let that create the output. It will resolve the clash according to the output specification.
2. Define two communicating processes, one for input and one for output. The output should be drawn to be compatible with the LDM structure. A separate formatting or sorting process can be created, as an output process, to handle the output.

INTERLEAVING CLASH

This happens in one data structure, the input. The data elements in the input structure represent more than one entity, but all are interleaved together, rather than all for the first entity, then all for the second and so on. To resolve this, Physical Process Design creates work space and extra processing to group the data elements in the optimum way for processing.

Common processing

During LDPD, it will be found that several processes share common processing, perhaps in input and output handling, sort/formatting, or database access. When such common processing is identified, a common module should be created and invoked. The Function Component Implementation Map in Stage 6 specifies how such common processes will be implemented physically.

Success units

When designing a logical database process the designer and User together must define the *success units* for each process, i.e., how much of the process must be performed for it to be deemed a success.

During the execution of a success unit, those portions of the database being accessed are effectively locked from other processes, so that the state of the database at the end is consistent with its state at the beginning of the process.

If anything should happen during the process to cause it to be abandoned—discovery of an integrity error, say, or the User who initiated an on-line task having to leave it for something else—the database will be 'rolled-back' to its state at the start of the process, cancelling any updates that were carried out up to that point. If the success unit is completed successfully, then the database is 'committed' or 'rolled-forward' to a new state. In this case, all updates will be effected on the new database.

The designer must agree the success unit strategy with the User. There are a number of possible strategies open to them:

An Update Process is a logical success unit

In this strategy—which is more or less inevitable—each Update Process will be a success unit. This is recognizing that an event has triggered the update, and that all changes resulting from this event must stand or fall together, rather than be implemented piecemeal.

An Enquiry Process is a logical success unit

This strategy states that an enquiry is triggered by a specific input, and while it is being executed the whole database should be locked to preserve integrity. Thus the trigger is equivalent to an event, and all accesses must stand or fall together, in a similar way.

Off-line functions will almost certainly be implemented this way; on-line functions may need to be broken down into smaller processes, or use extra modules to cope with interleaving clashes or boundary clashes.

An Enquiry Process may be divided into smaller units, where each is a logical success unit. This is a possible strategy for an on-line enquiry which accesses large amounts of data. The justification for this is to give the User a chance of abandoning the enquiry—for whatever reason—before the end, but without losing all the information gained up to that point.

It may be, for example, that an enquiry is to retrieve a piece of information that occurs somewhere in two large sets of data. The information may be an accumulated set of totals; it may be a single event that happened once, but when is not known. If the specific piece of information is found half-way through the enquiry, the User will not want to search through the rest of the possible records when there is nothing useful left to find. If the search is accumulating totals, say, the User will not want to lose everything done so far if the system should crash, or should the User be called away. In these cases, the logical success unit would be divided into chunks of the data accessed, say 10 records at a time.

An enquiry–event pair may be a logical success unit

If an Update Process involves making an enquiry first and, on return of data to the screen, then performing the update, the entire operation should be regarded as a success unit. If the two are made into separate success units, there is a danger that between the two the database will have been amended by another process. Thus an erroneous update will be made on the strength of outdated information.

During Function Definition, such enquiry–update pairs should be identified and grouped together. The role of the enquiry may be to provide the correct entry point to the database for the update, or it may prevent the User from updating the wrong occurrence of the entity. When such a pre-event enquiry is found, the database should be locked from the initial enquiry through to the end of the update.

Summary

Logical Database Process Design is the first of our design activities. Although Technical System Options have been selected, the design is still product independent.

There are two essential elements in the stage: producing Enquiry Process Models and producing Update Process Models. Both involve the use of structured diagramming techniques, and are built on SSADM products from Stage 3.

Enquiry Process Models are built up on Enquiry Access Paths, from Logical Data Modelling combined with Input/Output Structures from Function Definition.

Update Process Models are built up on Effect Correspondence Diagrams, from Event/Entity Modelling.

Both sets of models are input to Stage 6, Physical Design, to be transformed into code, if an application generator or 4GL is used, or into program specifications if a 3GL environment has been selected.

EXERCISE

Figure 17.13a shows part of the Logical Data Structure for an order processing system in a mail-order company. Customers place Orders for goods, each order comprising a Header and a number of Line Details.

There is a requirement to provide details of a given Customer's Order Header details, with the associated set of Order Lines, and Product, if any items are still not despatched within 21 days of the Order Date.

The EAP for the enquiry is given in Fig. 17.13b, as is the I/O Structure diagram for the enquiry function Outstanding Customer/Order Enquiry (Fig. 17.13c).

(a) Convert the EAP to a Jackson structure.
(b) Identify the correspondences between I/O Structure and the Jackson structure derived from the EAP. Which is input structure and which is output structure?
(c) Merge the two structures and form the Enquiry Process Model, including operations and conditions.

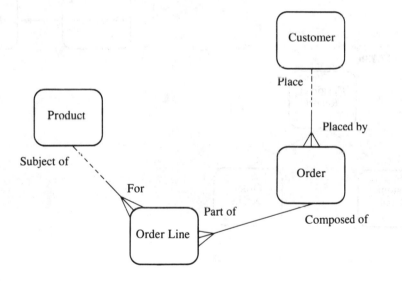

Figure 17.13a

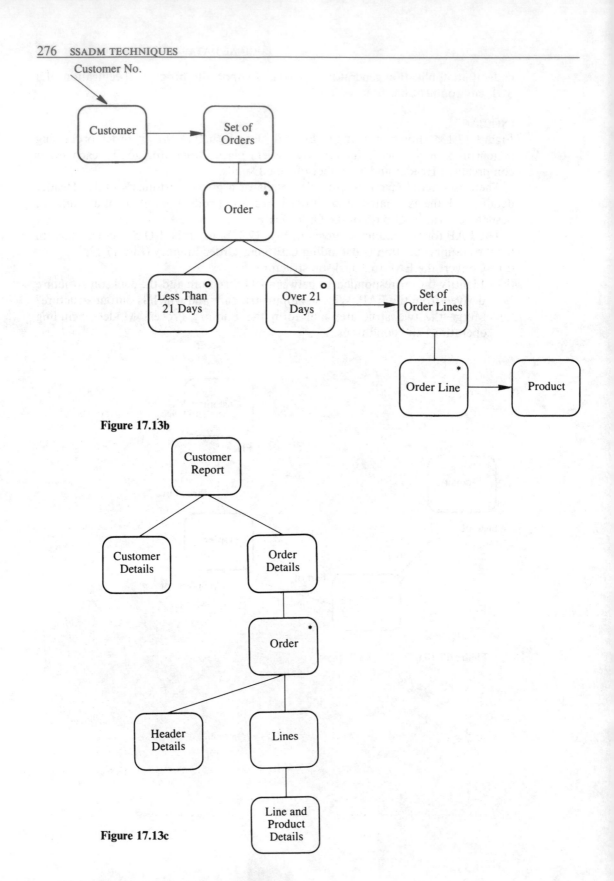

Figure 17.13b

Figure 17.13c

18. Physical Data Design

18.1 Aims of chapter

In this chapter you will learn:
- How to convert the Logical Data Model into a universal first-cut Physical Data Design
- How to classify the target implementation DBMS
- How to optimize the Physical Design to meet the performance objectives
- Some common methods for implementing relationship data

18.2 Where Physical Data Design is used in SSADM

Physical Data Design is carried out in three steps.

Step 610—Prepare for Physical Design The designer must study the target DBMS and classify its storage and retrieval characteristics. There are two forms to be completed in this task, the DBMS Data Storage Classification form, and the DBMS Performance Classification form. These should, where possible, be completed by the DBMS vendor.

Step 620—Create Physical Data Design A series of transformations is applied to the Logical Data Model to turn it into a universal first-cut data model. This model is product independent; to complete the transformation to a Physical Design we apply the rules peculiar to our selected DBMS. The particular facilities that we shall use in this design will have been selected in Step 610.

Step 640—Optimize Physical Data Design The Physical Design from Step 620 must be tested against the critical processes specified in Step 630. The aim is to achieve the performance objectives for the functions. Where this cannot be done, the design must be optimized in as many passes as are required until they are met.

Use in SSADM

The first activity in Physical Design Module is Physical Data Design. Because it is concerned with the placement of the data model on to any one of many DBMSs, it is not possible to describe the subject in more than generic terms.

The object of Physical Data Design is the placement of the new system data on to the storage medium, in such a way that the required accesses and updates can be made within the prescribed performance levels.

There are general principles that the designer must understand to achieve an efficient placement. As the principles are to do more with database technology and traditional file organization and design techniques than with SSADM, I shall not describe them in great detail. I shall indicate the areas in which efficiency can be achieved, especially in terms of navigating data hierarchies.

Although ideally the first-cut design should achieve the system's purposes, that is very rarely the case. After producing a product-specific design, using that product's own rules, we must carry out a paper timing exercise on the critical transactions, and where we find that the model does not support the performance levels, we must tune the design in various ways until we do achieve the required levels. If that cannot be done, then we must negotiate with the User to alter the levels to a target that we can achieve.

Inputs to Physical Data Design are as follows:
- Data Catalogue
- Effect Correspondence Diagrams
- Enquiry Access Paths
- Function Definitions
- Installation Development Standards
- Required System Logical Data Model
- Requirements Catalogue
- Technical Environment Description
- Update and Enquiry Process Models

Outputs from Physical Data Design are as follows:
- DBMS Data Storage Classification
- DBMS Performance Classification
- Function Definition (with sorting requirements)
- Physical Design strategy
- Requirements Catalogue
- Space Estimation form
- Timing Estimation form
- Physical Data Design

Figure 18.1 shows the relationship of Physical Data Design to other SSADM techniques.

18.3   Procedures

Physical Data Design is carried out through a series of activities. The three step titles summarize the broad approaches of the activities:
1. Prepare for Physical Design.
2. Create the Physical Data Design.
3. Meet performance objectives.

Prepare for Physical Design

Before we examine the Logical Data Model for the new system, and transform it into a Physical Design, we must first study and understand the implementation environment.

SSADM provides a means of classifying the chosen DBMS according to two basic considerations: data storage facilities and performance. There is a form for each that is to be completed. The DBMS vendor should, ideally, complete these forms, but it

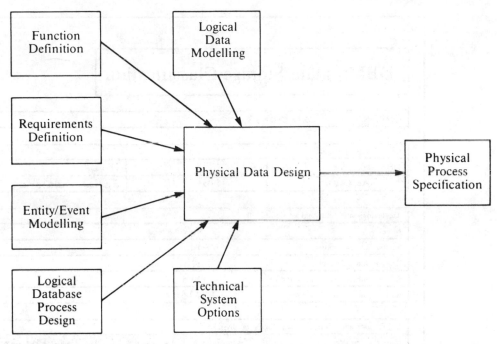

Figure 18.1

may well be that the designer must instead. The forms identify the particular mechanisms of storing and accessing physical groups of data that any **DBMS** may feature. In order to carry out the product-specific design, the designer must know and understand the particular mechanisms that the target DBMS employs. A discussion of these possible mechanisms is given later.

Figures 18.2 and 18.3 give examples for the two forms, using a typical relational environment. They are not connected to a specific product, such as DB2 or Ingres.

Once the classifications are complete, the designer must design a Timing Estimation form and a Space Estimation form. These will not be completed until after the product-specific design, but as their contents depend upon the classification that has just been carried out, so they are designed at this point. Examples of these forms will be given later in this section.

Once the implementation environment is thoroughly understood, and the designer is familiar with the particular storage and performance features available, the strategy for the rest of Physical Design can be agreed.

These decisions may include choosing which of the available facilities will be used, particularly those that deal with indexing and paging, for high performance. On the other hand, the decisions may be at a higher level, and may include such decisions as whether or not to move to a product-specific design, leaving out a universal first-cut design. Decisions like this will only be made by experienced designers, who are able to draw on that experience when carrying out the design.

All decisions made during this process should be open to revision right through the design stage, rather than be fixed and committed forever, once made.

DBMS Data Storage Classification

DBMS/file handler: *Relational*

Relationship representation

Table	List	Phantom
Indexes	*No*	*Serial searches*

Amalgamation of entity and relationship data

None	*Yes*
Relationship and master	*No*
Relationship and detail	*No*
Relationship and with master and detail	*No*
Relationship and relationship	*No*

Key representation in relationship (logical or physical)

Master to detail/detail to next detail	*Physical (Indexes); Logical (Phantom)*
Detail to master	*Logical (TNF data)*

Retrieval by logical key

Search	Indexing	Hashing
Yes	*Yes*	*No*

Implementation of place near logic

Clustering Indexes

Significant restrictions

None

Figure 18.2

DBMS Performance Classification

DBMS/file handler: *Relational*

Transaction logging overhead

For standard transactions which commit data only at termination, log overhead is minimal, as this is the only synchronous log I/O. Frequent commits increase the overhead

Commit/backout overhead

Commit—see above. Back-out for on-line transactions, usually minimal, may be substantial in batch processes without frequent commits, because of the need to read active log datasets (disk) and possibly archive logs (usually cartridge)

Space management overhead

Tables can be set up with free space to improve performance, particularly if data is volatile. Requirements for each table-index should be assessed and appropriate sizing adjustments made.

Dialogue context save/restore overhead

In CICS dialogue context is only saved under specific application control — no automatic features.

Standard timing factors

Disc operation:	Time:	Comment:
Read:	20 ms	Assume 3380 disk
Write:	20 ms	
Overflow overhead:		

DBMS operation	DBMS CPU time	TP monitor CPU time

An average call uses approx 15-20 000 CPU instructions, i.e. about 1 ms elapsed on an IBM 3090-180 (15 mps). DB has elaborate mechanisms to reduce syncronous I/O, especially for read only applications. Ratio of I/O requests to actual physical I/O should be about 5:1. CICS CPU time is typically minimal (per transaction).

Performance parameters for available sort packages

Figure 18.3

Create the Physical Data Design

This activity describes the first-cut design, i.e., a design has been produced following data design rules and guidelines, but it has not yet been tested against the prescribed performance levels. As a result of this testing the design will probably need tuning until the service levels can be met. Different methods and options for tuning the design are discussed later in this section.

The objective of this activity is to create a Physical Design that matches the rules of the chosen DBMS, that can be created swiftly, and against which the required processes and enquiries can be measured.

Before we look at the algorithm for transforming the Logical Data Model into a first-cut design, we must recognize certain assumptions made by SSADM, on which the algorithm is based:

- Entities on the LDM are represented as *record types*. Each occurrence of the entity is accessed as a record.
- Records are stored on *blocks*. The block (or 'page') is the physical unit of transfer. Each record is stored on an identifiable block, so that it can be accessed. The access is through the primary key, if a direct access is required, such as a Scheduled Course, or by being stored on the same block as a record which is accessed directly. An example of this would be storing Bookings records on the same block as the Scheduled Course to which they relate.
- Records are *grouped* into physical groups. Records or record types that are accessed together need to be grouped together on the disk. An occurrence of a detail record, therefore, must be stored in physical proximity to its master. Such groupings should be identified in the first instance on the LDS, as it is being transformed. The data design is built around the grouping on the same block of masters and details, and details of details, etc., according to the frequency of access. The actual mechanism for so storing and retrieving the data depends on the facilities offered by the specific DBMS.
- *Primary relationships* within a physical group are supported. Primary relationships are those between master and details in the same physical grouping. The DBMS will support these relationships.
- *Secondary relationships* between physical groups are supported. Relationships between entities from different physical groups are known as secondary relationships. These are supported, but the mechanisms for this form of storage may be different from those used by primary relationships.

With these assumptions in mind, we can now begin to transform the LDM for the required system into a first-cut Physical Design. This happens in eight steps.

Step 1—Identify features of the Required System LDM required for Physical Data Design This involves creating another model from the LDS, but omitting all information not needed for Physical Design. This model is only a working document, not the final design, nor is it a replacement for the LDS.

This is not a mandatory step: experienced designers who are fully conversant with the DBMS may not require this model. If a software tool is available, this would be an appropriate task for it.

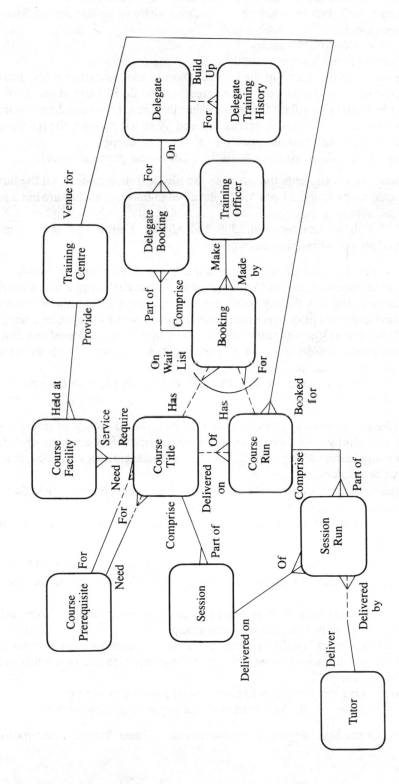

Figure 18.4

To achieve the transformation, apply the following guidelines:

- Turn each 'soft' box to a 'hard box', i.e., one with square corners, to distinguish the two models.
- Remove all relationship names.
- Replace all dotted lines with solid lines.
- Each relationship which, on the Logical Data Model, is dotted at the detail end (showing that the master need not be present), should be marked on the physical model by putting an 'o' on the relationship line, just above the detail entity.
- If a master has two exclusive details, marked by an arc crossing the relationships, replace the arc with two optional 'o's on the relationships.
- Include all the design volumes from the LDS on the physical model.

This model now represents the basis for the physical data design. All the business rules reflected in the Logical Data Model have been removed, as they are not a part of the physical storage plan.

Figure 18.4 shows the required LDS and SS plc. Figure 18.5 shows the LDS transformed into this first physical model.

Step 2—Identify the required entry points and distinguish those that are non-key   The entry points are derived from the set of Enquiry Access Paths and Effect Correspondence Diagrams. Identify the key access points on the model by a small circle with an arrow pointing to the entity concerned, and annotate with the key data item.

Identify the non-key access points on the model by a lozenge-shaped box linked by a crow's-foot relationship to the entity concerned, and annotate with the names of the (non-key) data items that access the entity.

Figure 18.6 illustrates the Physical Data Model for SS plc with the access points marked.

Step 3—Identify the roots of physical groups   The physical groups of records will be viewed as a hierarchy. The topmost entity in the hierarchy is the *root* entity. The first and most easily identified root entities are the reference entities, i.e., those that have details, but no masters.

The other root entities are those identified as direct access points, unless that entity has a master already identified as a root entity.

All entities identified as root entities are marked with a stripe along the top, as shown in Fig. 18.7.

Step 4—Identify the allowable physical groups for each non-root entity   All non-root entities should be put into a physical group, according to one of the following constraints:

- A non-root entity may only be put into a physical group if one of its mandatory masters has already been put into that group.
- If a non-root entity is a direct access point, and it has more than one mandatory master across different groups, put it into the group with the master whose key is part of the entity's key.

Draw lines around the groupings to show all the possible groupings.

Figure 18.8 shows the SS plc model with the physical groups marked.

Step 5—Apply the least dependent occurrence rule   Under Step 4, a non-root entity

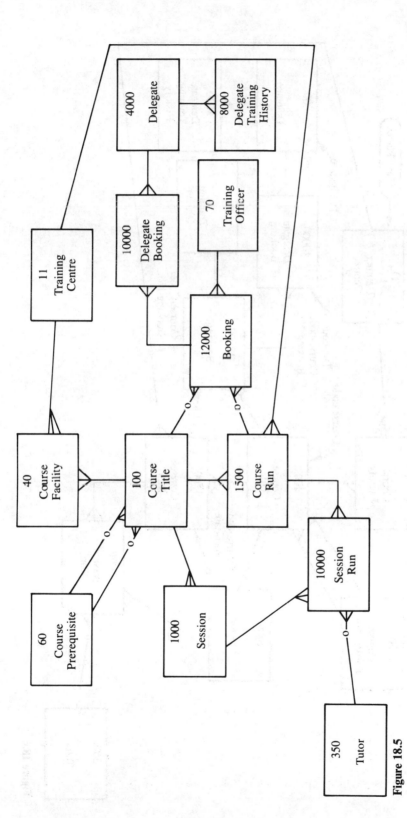

Figure 18.5

285

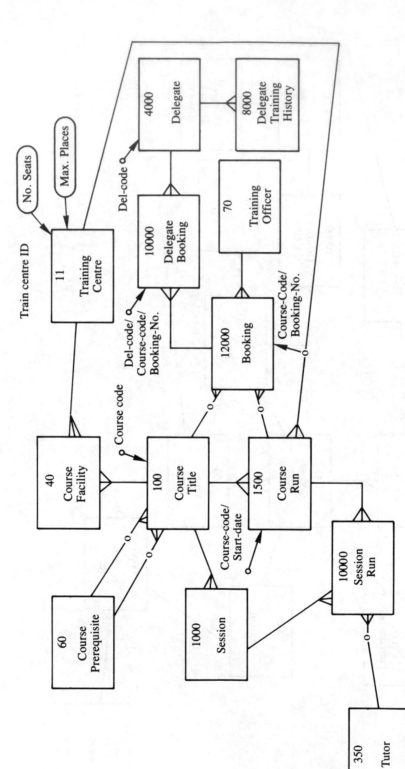

Figure 18.6

286

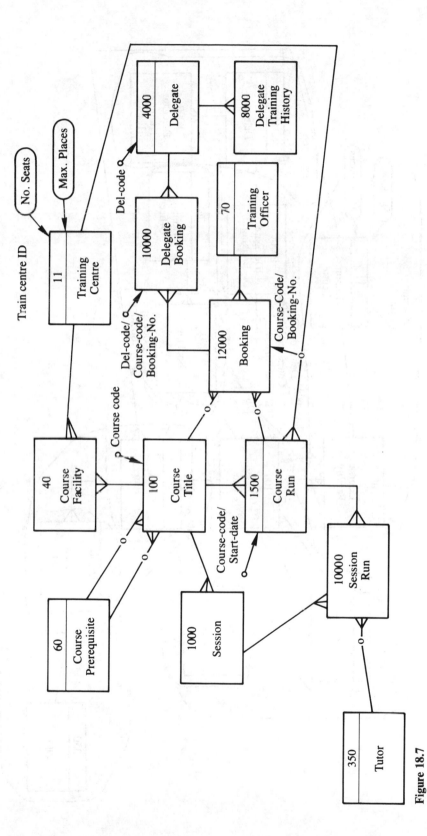

Figure 18.7

287

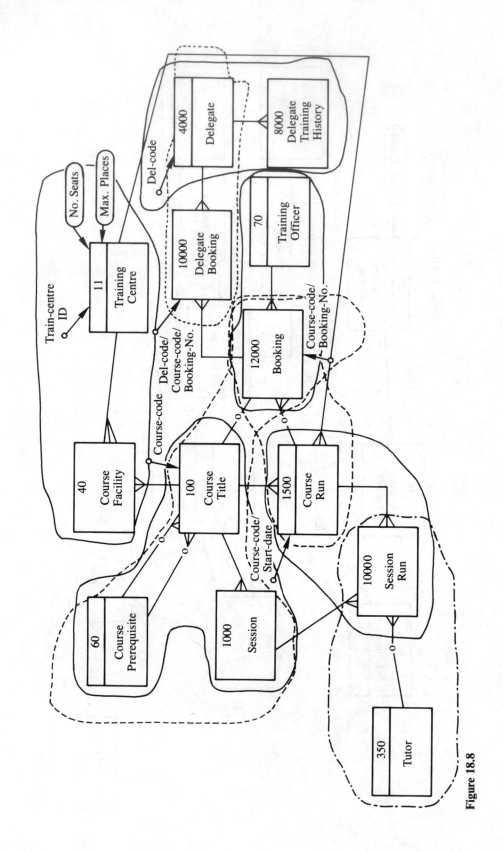

Figure 18.8

288

may find itself in several allowable physical groups. As it may only be put into one actual group, we apply the following measure to allocate it to just one: when an entity may be put into more than one physical hierarchy, put it into the one where it will have the least number of occurrences. This is not an infallible and inevitable rule, but a generally reliable guide.

Figure 18.9 shows the model for SS plc with the least dependent occurrence rule applied.

Step 6—Determine the block size to be used The aim of this step is to select a block size that supports the largest of the commonly used blocks. The information can be obtained from the Function Definitions. The chosen size must be supported by the target DBMS, and must not cause buffer problems when read into memory, especially when several transactions arrive close together.

When choosing a block size, therefore, take the following considerations into account:
- The block sizes that the DBMS will support.
- The size of the most commonly used physical groups.
- The amount of space that the blocks will take up when they are read into memory.
For our current example, I shall select a block size of 40K.

Step 7 Split physical groups to fit chosen block size Once the block size is known, the physical groups must be measured against it, to be sure that each block fits. If it does not, the groups must be split in such a way that they do fit.

There are two calculations to make: the size of the data in the groups, and the storage needed for relationship data.
- To calculate the size of data storage, add the lengths of each data item for each record type and multiply by the expected number records of that type.
- For calculating space for maintaining relationships and disk management, sizing standards should be available. If they are not, the following rules of thumb may help:
  —For each relationship, allow six bytes in the master, and six bytes in each detail.
  —If there is to be a secondary index, allow six bytes per entry.
  —Allow six bytes per record for header overheads.
  —Allow between 30 and 40 bytes per block for block overheads. Assume that there will be a block packing density of about 60 per cent to reduce overflow.

Enter these estimates on a Space Estimation form. Remember that at this point these are first-cut estimates only, and will be revised considerably. Their function at the moment is to enable us to see whether our physical groups will actually fit into the blocks. One of the Space Estimation forms for SS plc is to be found in Fig. 18.10.

If a group does not fit, then we must split it. We do this from the bottom of the hierarchy upwards, to find a subgroup that will fit into a block. When splitting groups in this way, re-apply the least dependent occurrence rule, If you have to do this, then complete a fresh Space Estimation form for the new grouping. Each time you have to revisit a data group in order to fit it into the specified block size, you must update the relevant documentation. (You do not need to hang on to the superseded forms; only the current valid ones are part of the deliverables.)

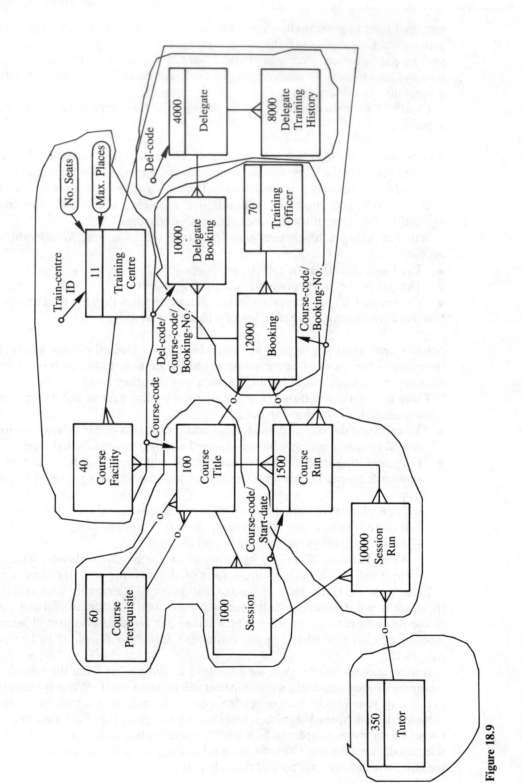

Figure 18.9

Space Estimation Form

| Block size: | 40K | | Packing density: | 60% | 24K | | Range from: | 1 | To: | 40 |

Record	Key	Records per hier-archy	Data size	Record o/head	Primary index	Second-dary index	Space per record	Record space per block
Course Title	Course Code	1	100	36	6	18	160	160
Session	Course/Session Code	10	100	18	12	–	130	1300
Manager	Duty Code	1	50	18	6	6	80	80

Data total	1540
Block header	40
Block total	1580

Note: There is room for approx. 21 hierarchies per block;
allow load space, and security requirements
and we can still load 15 hierarchies.

Figure 18.10

Step 8—Apply the product-specific rules to the design   Following the DBMS classification exercise we know about the various facilities offered by the DBMS. Now we must choose which of them to apply to data design. Once we have chosen them we convert our universal first-cut design to a product-specific design, based upon the design rules of our target DBMS.

These rules are provided by the vendor. It may be that some products do not include such rules; in this case, the rules must be created.

In deciding upon the features to be included, and in applying the rules, there are three issues to be addressed:

- *How are 'place near' mechanisms implemented?*—We must store the physical groups on the data model as collections of records in as close proximity as possible on the same block.

  Each DBMS has its own way of storing such records or, if they must be separated, of preserving the link between them for fast access. Some DBMSs are more flexible than others in providing for this feature.

  It is up to the designer to choose the most appropriate mechanism for implementing this feature, and to what extent it will be used.

- *What is the support for secondary relationships?*—The place-near facility affects records in primary relationships, i.e., those related to other records in the same physical group. This facility should also be made available for secondary relationships, i.e., those records that are related but are in different physical groups. In such cases, the DBMS should provide an efficient way of moving from master to detail, or from detail to master, across two physical groups. There are some DBMSs that do not cater for this facility; in such cases the designer must build one instead.

- *What are the specific restrictions imposed by the product?*—This heading can apply to almost any aspect of data storage, but in practice it usually applies to just two areas:

  —Restrictions on how data on relationships can be amalgamated with data on entities. This usually applies to recursive or optional relationships.

    If there are such restrictions, the designer must find a way of representing the data in spite of those relationships. Ways of doing this include creating secondary indexes, creating special records, or fields in records, to record the relationships.

  —Restrictions on how records can be placed near other records. If restrictions are imposed here, the designer must reduce the extent to which records are blocked together, by splitting the physical groups into smaller groups.

Meet performance objectives

After we have completed the first-cut Data Design, we prepare the Process Design in Step 630. Having defined the processes, we can now, in Step 640, apply them to our product-specific model, timing them to see if the prescribed performance objectives can be met.

The timing and storage constraints were identified and recorded in the Requirements Catalogue during Stage 3. If the product-specific design meets those constraints straight off, then it will be implemented as it is. This is rare, however, and more usually the design must be tuned, or *optimized* until it does meet the requirements.

There are two broad approaches to this: if there is a tool available locally to generate a database easily, with similar characteristics to the target environment, the timings can be tried using benchmarks. If that is the case, then SSADM guidelines need not apply.

If no such tool is available, the following SSADM considerations and activities will enable the designer to optimize the design to meet the objectives as far as it is practicable.

Optimization is not a quick and easy task, and carries overheads in the implemented design, so any such exercise should be carefully planned, with the objectives understood before it is commenced. It should be carried out by an expert in the target environment, and, where the payoffs no longer seem to justify the effort, should be stopped, and the performance objectives renegotiated.

Figure 18.11 illustrates the optimization cycle and the main activities involved. One important activity not shown in the diagram is the consultation with the User over changing the requirements, rather than continuing in the cycle until it becomes counter-productive.

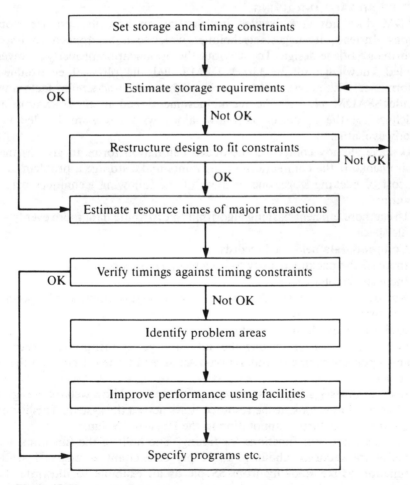

Figure 18.11

OBJECTIVES

If we are to optimize the design, we must try to achieve one of two aims:

- To save storage space
- To save processing time

Whichever is our objective, to achieve the one will adversely impact upon the other. In saving space we may increase processing effort, and in saving processing effort, especially to retrieve data, we may increase data storage. In either event, we are in danger of making the system rather more complex than our initial design. The danger here is that the system is less like the data model, and so less portable.

Where possible, our optimization should allow us to preserve the data model in our system, so that it will be easily maintained and be more easily understood. However, there are, more often than not, times when the model cannot be maintained; this may be because of limitations in the implementation mechanisms, or because the performance of the model is genuinely too slow for our requirements.

OPTIMIZE STORAGE AND TIMING

SSADM does not give detailed instructions on how to optimize storage and timings. Instead, it suggests techniques which can be applied to improve the performance of the design. To carry out the optimization properly involves a more detailed knowledge of the target DBMS and the physical environment, CPU performance, disk access speeds, access mechanisms, and similar factors which are outside SSADM *per se*. The techniques mentioned in these sections represent guidelines for the optimizer to follow, along with the more detailed knowledge already available.

We have already completed the Space Estimation forms to give us the storage requirements. If the configuration constraints make storage a problem, despite the low cost of backing store, one or more of the following techniques may help the situation:

- Use a more efficient hashing algorithm to distribute data more evenly across the file space.
- Compress data fields and records.
- Increase the packing density to hold more records per block.
- Increase the block size.
- Remove redundant data from master–detail relationships, and let pointers maintain the relationship.
- Reduce/remove historical data.

The problem with optimizing the data storage, as mentioned earlier, is that it impacts upon the timing considerations. An agreed trade-off between the two must take place.

As well as investigating the storage requirements and how well our design supports them, we must also look at the resource times across the system. This is not a small task, and involves further annotation to the Physical Design:

- Identify the major functions for timing. The bulk of the functions will not be major transactions; about 10 to 15 per cent count as major. I shall use the function Make Booking from SS plc as an example to illustrate the timing exercise.

- Time the transaction. This involves completing the Timing Estimation form we designed after classifying the system. It is completed by recording the following:
  —Physical access information. In this we include the record types that need to be accessed for the function, the numbers of each record accessed, and the access paths followed.
  —The disk accesses required to make the physical reads and writes above. If all the details are stored by a master, there may be several physical reads through the set until the target detail is found, but only the one disk access is needed to read the whole set in with the master accessed.
  —Complete the CPU time and the totals. For this we need the information we have already collected on the DBMS Performance Classification form. To complete this information, we must have the CPU figures made available to us, the translation of machine instructions through various layers of software, such as operating system, TP monitor, etc., and any algorithms or calculations for executing interrupts or scheduling.

Figure 18.12 shows the Timing Estimation form filled in for the function Make Booking.

- Identify problem areas. Having completed the timing objectives, look again at the performance objectives, and identify significant mismatches. These must be addressed and ways of improving performance devised, either by retaining the structure of the model, or by compromising the structure. In the first instance, we can use the DBMS facilities as identified, and in the second we may ignore them. I shall not describe any of the methods in detail, but indicate particular areas that the optimizer can address.

Improve performance by retaining the structure of the design and using the DBMS facilities
There are several techniques available to improve performance, all to do with storing access information on entities, on relationships, or on keys.

Entity The storage of entities can be optimized by any one of the following mechanisms:
- Place details near the masters whose relationship is most often used. Thus when a block with the master is read in, the details it requires most will also be read in.
- Change the access method, so that hashed keys become indexed, or indexed keys become hashed. Obviously, questions such as 'what is the hit rate for this record set going to be for the major functions?' must be asked when deciding to do this.
- Implement key-only entities as indexes, to allow a more rapid navigation.
- Add direct access mechanisms to detail records, so that physical disk access does not entail reading such a large hierarchy of records before finding the target.

Relationship Add extra pointers, backwards and owner pointers especially.

Key Read the entire pointer set into memory at the beginning of a function, and store in a look-up table. This can prevent a costly multipass search through sets to retrieve a relatively low number of records. This minimizes the overhead of DBMS to CPU.

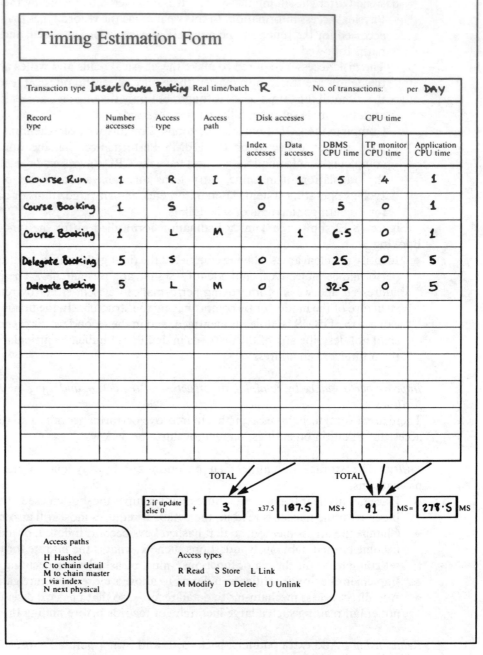

Timing Estimation Form

Transaction type *Insert Course Booking*   Real time/batch **R**   No. of transactions: per **DAY**

Record type	Number accesses	Access type	Access path	Disk accesses		CPU time		
				Index accesses	Data accesses	DBMS CPU time	TP monitor CPU time	Application CPU time
Course Run	1	R	I	1	1	5	4	1
Course Booking	1	S		0		5	0	1
Course Booking	1	L	M		1	6.5	0	1
Delegate Booking	5	S		0		25	0	5
Delegate Booking	5	L	M	0		32.5	0	5

TOTAL   TOTAL

| 2 if update else 0 | + | **3** | x37.5 | **187.5** MS+ | **91** MS= | **278.5** MS |

Access paths

H Hashed
C to chain detail
M to chain master
I via index
N next physical

Access types

R Read   S Store   L Link
M Modify   D Delete   U Unlink

Figure 18.12

There are various other ways of tuning the physical design that do not affect the one-to-one mapping. Common ways include:

1. Adjust the block size to hold more records in a single block.
2. Increase the number of blocks held in buffer.
3. Increase the packing density in the blocks; before going for this option, examine the likely volatility of the file first, and the impact of frequent overflow.
4. Store the most active data on faster devices.
5. Optimize the disk placement of files. Where possible, it is more desirable to choose an optimization path from this section, rather than have to compromise the design. Departing from the logical model may improve performance, but it also makes maintenance of the design more difficult. Another danger in compromising the design is that *ad hoc* query facilities are also compromised. If the facility for such enquiries were a major requirement, then we have to forfeit a major requirement for performance in another part of the system.

Before carrying out any tuning, check with the User whether or not the performance levels achieved are acceptable. Even if the timings do not reach the specified levels, the User may accept them, and so tuning is superfluous.

Improve performance by compromising the structure of the model or by not using DBMS facilities

If tuning is required, and the required levels cannot be reached by preserving the integrity of the model, then one or more of the following techniques may do the trick. Again, I do not propose to describe the mechanics of each in detail, but will just indicate areas where action may be considered.

Redundancy
1. Incorporate derived data, such as results of calculations, in the masters to save processing time.
2. Incorporate data belonging to the master record in the details.

Entity representation
1. Merge details into masters (incorporate repeating groups into the master).
2. Split the logical entity into two or more record types, related by key or by set.

Relationship representation   Choose another way of representing the relationship other than by a set. Possible means include indexes, pointer arrays, or pointer chains.

Delay updates
1. Postpone all updates to records until after the transaction is completed: they can be carried out in one pass later.
2. Postpone all updates which change relationships.
3. Postpone all updates which delete records.

In all cases the records would be marked so that logically they had been changed, but the physical activity would not take place until a subsequent occasion.

It is important to remember that tuning a design to meet the required performance levels is a costly task, and should be undertaken in a pragmatic manner. If the result is too far from the original model, or if the task of tuning is going to take an unreasonable time, then negotiations with the User should be started with a view to altering the performance objectives.

As with the Space Estimation form, complete a fresh Timing Estimation form for each optimization path. If, for example, you choose to store master data inside detail records, there will be fewer accesses and fewer reads. The new form must reflect the difference in timing. Similarly, if you change from indexed access to a hashed access, there will be at least one less disk access, maybe more, and so you must complete a new form to show the timings on this path.

Complete the Physical Data Design

After tuning is complete, and designer and User are content with the results, the remaining tasks for Physical Design are completed.

Validate the impact of imposed sequencing   Examine the impact of any sorting of transactions. A sort means that the model will be hit by the transactions in a different order to that in which they occurred in the world. This may or may not present a problem. To check the validity of this new order of events, look again at the Function Definitions investigations for the function in question.

Document the sorting requirements   Where sorts are to be included, be sure that they are entered into the Function Definitions.

Identify processing optimization requirements   If the data optimizations carried out previously do not meet the performance objectives, then consider optimizing the code. This may be a matter of writing more efficient code in the same implementation language, or, if an application generator was used, perhaps some code written more efficiently in a 3GL may be needed; perhaps it may even be necessary to code a module or routine in an assembler language to achieve the required performance.

If such requirements are identified, they must be documented in the Requirements Catalogue.

Update the service-level requirements   If the performance objectives cannot be met, or if two separate functions demand conflicting objectives, the designer and User must agree on new service-level requirements. Where this happens, the Requirements Catalogue and the relevant Function Definitions must be updated accordingly.

Record optimization decisions   Any complex optimizations should be recorded for future maintenance or enhancements. SSADM does not provide a product specifically for this, but the design documentation should cater for such optimizations.

Validate the performance of the final design   Only the critical functions will be tested here. First build the Process Data Interface (PDI) Modules. These are explained in Chapter 19 (Physical Process Specification) in more detail. If the environment exists on site, they are better built using a non-procedural language. This is preferable, but not essential.

Build dummy Enquiry and Update Modules and submit them to a heavy series of test transactions. Monitor the performance of the database against these tests for any errors or miscalculations in the timings. If the results are satisfactory, the PDI Modules can then be used in the final implementation.

Summary

Physical Data Design is impossible to prescribe in strictly algorithmic terms, because the variety of implementation and development environments is so wide. If there were a standard to which each manufacturer and supplier worked, this subject could be more clearly defined. Nevertheless, SSADM defines the task as closely as possible in a generic way, always keeping the central tasks in mind: to implement the logical model as closely as is feasible, and to implement it such that performance requirements are met as closely as is feasible.

The specific requirements for data design and the hooks for SSADM are expected to be supplied by the vendor, rather than by SSADM or CCTA.

After the product-specific first-cut design, the designer enters into an optimization cycle that continues either until all performance objectives are met, or until it is counter-productive to continue. In the latter case, the User's agreement to change the objectives must be obtained.

19. Physical Process Specification

19.1 Aims of chapter

In this chapter you will learn:
- The purpose of Physical Process Specification
- How to classify the Physical Processing System
- The defining of a Physical Design Strategy
- When to create the Function Component Implementation Map
- When to prepare and consolidate the Process Data Interface

19.2 Where Physical Process Specification is used in SSADM

Physical Process Specification is carried out in the following four steps:

Step 610—Prepare for Physical Design   In this step we must study the selected physical environment to establish how the Logical Design will be implemented. The decisions made here are all product specific.

Step 630—Create the Function Component Implementation Map (FCIM)   We break each function down into its component parts. We examine each function to define its implementation plan, and to identify any fragments that can be reused.

As well as specifying for each the handling of the processes and data streams so far unspecified, we identify common processing across the Modules and in so doing remove duplication.

The end of this step is a definition of database access paths, a documented Physical Design Strategy, and specifications for reusable processing fragments.

Step 650—Complete Function Specification   All specifications for functions that require procedural code are defined here. Specific Function Models can be drawn for any where there are divergences from the Universal Function Model. Structure clashes are resolved, and any recognized optimization requirements are specified.

Batch programs and procedural and non-procedural components will be sequenced together, along with any sort or other utilities required.

Step 660—Consolidate Process Data Interface   The fragments on the FCIM that access data are compared with the optimized Physical Data Design. Where differences are identified, extra fragments, or Modules, are specified to act as a mask for these mismatches. If a non-procedural language is available at the installation, the mask, or interface, can be written in that language at Step 660. If there is only a procedural tool available, the intervening Module will be specified here, for subsequent coding.

Use in SSADM

The purpose of Physical Process Specification is to translate the logical specification into physical programs, I/O formats and Dialogue Designs. These products will be designed for a specific environment.

Inputs to Physical Process Specification are as follows:

- Logical System Specification
- Installation Development Standards
- Physical Environment Specification
- Application Style Guide

Outputs from Physical Process Specification are as follows:

- Application Development Standards
- Function Component Implementation Map
- Function Definitions

Figure 19.1 shows the relationship between Physical Process Specification and other SSADM techniques.

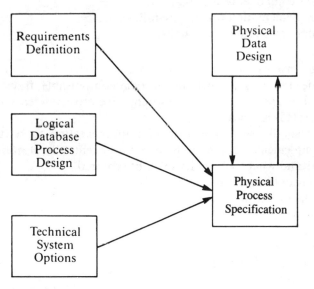

Figure 19.1

19.3 Concepts of Physical Process Specification

Physical Process Specification makes use of several techniques that have not been used before, and creates new products as intermediate stages to the production of Program Specifications, or Modules.

Before describing the activities of the four steps in greater detail, I shall describe the new products and concepts used in Physical Process Specification.

Fragments

'Fragments' is not a name for a new technique, but it is a term that will be used frequently. It describes a defined piece of processing that may be at any level. Components of functions may themselves be broken down into fragments. A fragment must be defined in terms of purpose, of inputs and outputs and correspondences. It may be at a level with operations, or it may be to do with displaying a screen message or defining a data group. The Function Component Implementation Map (see below) must define all processing at fragment level, as well as at function level.

Function Component Implementation Map (FCIM)

PURPOSE OF THE FCIM

Much of the effort in Physical Process Specification goes into compiling the FCIM. The FCIM's purpose is to classify and specify all implementation fragments for all components. It is a network diagram that shows how the elements of the processing in Physical Design fit together, and how the logical function components map onto the physical.

The FCIM has four aims:
1. To eliminate duplicate components and fragments.
2. To reuse common components and fragments.
3. To specify the implementation route to be followed.
4. To package components into success units.

COMPOSITION OF THE FCIM

The FCIM is made up of specifications of common/reusable fragments and components. As well as specifications, a comprehensive cross-reference of related functions and fragments is produced.

The specifications should show the form of implementation to be used, i.e., whether it is to be implemented procedurally or non-procedurally; if non-procedurally, what languages and what tool facilities will be required.

Each specification should describe:
*   The purpose of the procedure
*   Its relationship with other procedures
*   Inputs and outputs
*   Summary of operations

Programs, or run units, will also need:
*   Constituent programs
*   Hardware requirements
*   Files:
    —Media
    —Names
    —Sort order
    —Volumes
    —Volatility
    —Record layouts
*   Transform operations

- Controls
- Checkpoint/restart requirements

The descriptions are consistent with the specific implementation environment and data dictionary conventions.

Manual procedures, too, are defined as part of the FCIM. Features described include:

- Purpose of the procedure
- Relationships with other procedures
- Resource requirements
- Management policies that impact on the procedures
- Timings

Non-procedural specifications

There are two types of implementation path for each function component: it may be coded in a procedural language, or it may be encoded in a non-procedural language.

Where possible, a non-procedural implementation should be selected. This path leads to fewer errors, and hence lower maintenance overheads. Development times will probably be lower, and greater reuse of non-procedural fragments is likely.

SELECTING NON-PROCEDURAL IMPLEMENTATION PATHS

Data descriptions are always coded non-procedurally, whether the description is of a file structure, a screen layout, or a report format.

If a system is simple, rather than complex, i.e., it has clearly defined access criteria and minimal processing or transformation, a non-procedural implementation is recommended. The complexity of such systems can be deduced by examining the Entity Life Histories: if the ELHs are trivial, the system will be fairly simple, however large the data model.

If a non-procedural language is available, it should be used to specify the Process Data Interface (PDI).

ALTERNATIVES TO NON-PROCEDURAL IMPLEMENTATION PATHS

If the processing is inappropriate for a non-procedural path, or if the installation/ environment does not support a non-procedural tool, then the process must be defined in a procedural language.

Some notation for designing the algorithm must be made, according to the Application Style Guide. To achieve consistency, the structure diagram method used to devise the LDPDs would be completely appropriate. Defining Input/Output Structures and merging them to form a process structure, and allocating operations and conditions means that, after Stage 6, all that remains to be done is to code the functions.

Often, procedural environments do include opportunities for some specification of non-procedural elements. Where this is the case, the opportunity should be taken and such elements described in the component specification.

Process Data Interface

The Logical Data Model is unlikely to be mapped exactly onto the optimized Physical Data Model. The processing requirements are similarly unlikely to map exactly onto the Logical Data Model. Where such a one-to-one match does exist, the process may be implemented without problem; where there is a mismatch, we wish to preserve the logical view of the data, while recognizing that the processes will not fit onto it. An intervening Module, the Process Data Interface (PDI), must be specified.

USE OF THE PDI

The PDI acts as a mask for the Physical Data Design, so that to the process components it appears as the Logical Data Model. This eases maintenance over-heads, and simplifies change as well. The Module interprets accesses specified to the database in terms of the LDM so that the correct physical data items are retrieved.

COMPOSITION OF THE PDI

The PDI is made up of FCIM fragment specification for DBMS accesses, and for control utilities and non-procedural syntax.

IMPLEMENTING THE PDI

Where possible, the PDI should be implemented in a non-procedural language. This entails defining a set of 'logical views' of the database. These views are made up of:
- The required operation on the LDM
- The requisite operation on the physical database
- Non-procedural specification of the access paths

If there is no facility for defining these non-procedurally, then the PDI should be specified as a set of procedural Modules that enable access to all required physical records for each operation.

19.4   Procedures

Prepare for Physical Process Design

Step 610 is now a product-specific step. The development and implementation environ-ments have been identified, and their facilities and constraints must also be identified.

In this step we carry out the following four activities:
1. Study the implementation environment.
2. Produce the Processing System Classification.
3. Specify the Application Development Standards.
4. Develop the Physical Design Strategy.

STUDY THE IMPLEMENTATION ENVIRONMENT

The physical environment has a number of features and tools associated with it. Before we can proceed with the design or specification, we must identify the features and facilities now available to us. The Physical Environment Description is a key input to this activity.

PRODUCE THE PROCESSING SYSTEM CLASSIFICATION

There is a form for the Designer to complete that classifies the facilities offered by the physical environment, as illustrated in Fig. 19.2. There are a number of issues associated with this form. To complete it successfully, the designer should be supplied with a product-specific guide from the environment vendor.

Processing System Classification

Classes of tool feature *Name of tools in physical environment*

Procedural/non-procedural ☐ Tick if both can be provided   Online/offline ☐ Tick if both can be mixed

Success unit

Define the alternative commit strategies that can be employed.

Error handling

Describe the flexibility and recovery contingencies.

Process components

Describe the ways in which processes can be combined and generated at run time.

Database processing

Describe DB access flexibility.

Update ☐   Enquiry ☐

I/O processing

How can different data items for different groups be related together on screen.

Dialogue processing

Describe which dialogue components can be generated.

Process Data Interface

Describe the extent to which a PDI can be generated non-procedurally.

Distributed systems

Figure 19.2

Particular issues that the designer needs to resolve connected with this form come under the broad headings of:

- Flexibility
- Generation of dialogue processes
- Generation of off-line processes
- Generation of physical database processes
- Automation of success units
- Automation of integrity errors
- Modularity of procedural processes
- Generation of PDI
- Distributed systems

Flexibility

The questions to examine here concern the mixing of procedural and non-procedural specifications in the one function. Similarly, can on-line and off-line processing be mixed in the same function?

In both cases, the answer depends on the facilities provided by the environment.

Generation of dialogue processes

Questions relating to dialogue specification cover the navigation and grouping of dialogue elements, and the mapping of physical data items/groups to displayed dialogue items:

- Can data items on screen be related from different groups?
- Can logical data groups be implemented directly as physical data groups?
- Can an error-handling dialogue component be generated?
- Can a dialogue component be generated to search for and select a specific entity?
- Can use be made of standard navigation mechanisms?
- Can these mechanisms be tailored?
- Can the dialogue navigation be specified in other ways?

Frequently, the designer will want the answers to a subset of these questions, rather than needing all facilities.

Generation of the off-line processes

Can the I/O processes of an off-line function be generated without generating physical database processing at the same time?

Generation of physical database processes

This issue examines the facility to generate database processes automatically, using particular features of the DBMS. If such a facility exists, it may not be suitable for the designer's purposes; alternatively, it may provide precisely what is required.

- Can physical update and enquiry processes be generated?
- Can any physical database processes that have been generated be modified or replaced by procedural code?
- Can physical database processes be generated without incurring performance overheads?

Automation of success units

Can the physical success unit be matched with the scope of a logical success unit?

Automation of integrity errors
Can the physical environment be prevented from automatically inserting integrity checks into database programs?

Modularity of procedural processes
Processes, defined logically as discrete processes, may require to be combined to run together as host/subroutine, or as shared subroutines. Questions regarding the facilities for combining such routines are:
- Can two processes be combined as host and subroutine?
- Can a process be made a subroutine of any other process?
- Can processes be combined as co-processes, and how?
- Can the data linkage between Modules be made explicit?
- Can single copies of compiled processes be kept and combined with other processes at run time only?

Generation of PDI
Are there facilities for specifying the PDI in non-procedural language?

Distributed systems
This consideration is obviously of specialist application only. The issue is whether the distribution of a database over several locations can be ignored.

SPECIFY THE APPLICATION DEVELOPMENT STANDARDS
The logical installation and the physical environment may both impose naming standards for each fragment. The standards can be supported by careful cataloguing of the fragments. This can help both in identifying potential reuse, and in maintenance issues.

The support can consist of an indexing/classifying description:
- ID (name)
- Class of fragment
- Purpose/action
- Inputs/subject/precondition
- Outputs/object/postcondition

This is a suggested description, but it is a useful way of expanding the information without extending the name length.

DEVELOP THE PHYSICAL DESIGN STRATEGY
The Physical Design Strategy (PDS) mostly involves making decisions about implementing the functions and fragments. The decision rests on whether they should be implemented procedurally or non-procedurally.

The criteria
The decisions are based on the criteria that are chosen and the facilities identified during the Processing System Classification. The classification form will enable the designer to answer two questions:
1. How much of the physical processing can/should be specified in a non-procedural language?

2. How far can the processes defined in the Logical Design be directly implemented as physical programs, modules, or subroutines, within the Physical Processing System?

Once the decision has been made for each function or fragment, the Activity Descriptions must be customized, and the FCIM standards defined.

Apply the criteria

Once the criteria have been selected for either procedural or non-procedural implementation, they must be applied. It may be that there is no choice in the matter, through lack of facilities in the environment.

If the choice does exist, however, then a decision must be made whether to implement database accesses dynamically, or in embedded code. The choice will depend on the trade-off between service-level requirements, such as performance, flexibility, and ease of maintenance.

As a rule of thumb it is best to implement update processes in a non-procedural way; the designer must be able to tailor or manipulate generated code, or to supplement it with procedural fragments.

Enquiry processes should be implemented non-procedurally. Most of them can be, with less ambiguity than update processes. The only question that affects enquiry processes: is the enquiry to be a success unit itself, or is it to be bound in with an update process as a pair?

Performance should be evaluated; if a process which has been marked down for non-procedural implementation is a critical process, and will run into performance problems, then perhaps it should be implemented procedurally, after all.

The following strategies can be implemented:

- Locate the database process code and modify it.
- Discard the database process, replacing it with procedural code.
- Implement both dialogue and database processes in procedural code.

Whichever strategy is selected for each function, it must be well documented, including the reasons for selecting that particular path.

Generate physical dialogues

There are three possible strategies for designing physical dialogues:

1. *Input and process one input at a time*—In this strategy, the natural way of processing would be to receive one input at a time. The designer must specify a dialogue to:
    (a) Present the target data group in such a way that only the one event can be input.
    (b) Process the one event.
    (c) Report the success or failure of the event.
2. *Input a batch of events and process each individually*   Here, the system recognizes the events input and processes them in sequence. The dialogue will:
    (a) Present the data group so that up to, say, three events can be input.
    (b) Detect the data items which have been changed.
    (c) Invoke up to three Update Process Modules, one after the other.
    (d) Continue until all have been processed, even if one or two should fail.
    (e) Display up to three success or error messages.

3. *Input a batch of events and process them together* The processing of this strategy is similar to that in 2, except that if one event should fail, everything will fail, including those already processed in the batch.

At the conclusion only one success message or error message will be displayed.

Define the FCIM standards

The components of the FCIM can be broken down into four distinct types, each of which should have specification standards defined for it. The four types, and their SSADM origin, are as follows:

Physical processes	*Logical origin*
I/O programs	Function Definition
Dialogue Control programs	Dialogue Design
Database programs	LDPD
Common Modules	Various

For each function, each of these types has a prepared Program Specification. Local Installation Style Guides to a large extent dictate the format of these specifications, but the design team must define the standards first.

Create the Function Component Implementation Map

The functions must be defined so fully now that the next task is to generate the code. If application generators are to be used in the environment, then the product of this step will be generated code. If it is to be a 3GL environment, the output will be specifications that can be passed to a programmer.

The designer is working on two levels during this step looking for duplication and commonality across all the functions at one level, and at the other specifying the FCIM elements as fully as possible. The two levels are described more fully below.

REMOVE DUPLICATION

The particular areas to look at here are Enquiry Processes. Duplication can be identified and removed by:

- Removing the enquiry from an enquiry–event pair. When an Enquiry Process is a part of such a pair, check whether the enquiry trigger is a subset of the event data. If so, discard it. Check whether the Enquiry Process is a part of the Update Process. If so, again, discard it.

  Thus, in SS plc's system, when making a Booking for a place on a course, the User retrieves details of that course to establish that there are places. However, the same retrieval is necessary to perform the Booking, whether or not the enquiry is to be made. In that case, the enquiry trigger is superfluous in processing terms, and can be omitted as a separate input.

- Creating an enquiry process by duplicating an Update Process. If the enquiry part of an enquiry–event pair has been created simply by duplicating part of the Update Process, without the operations, discard or change this. Enhancement and maintenance work is duplicated by this strategy.

If each part is a separate success unit, the update may be rejected because of invalid states, even though the enquiry part was accepted as valid.

If validation tests are included only in the enquiry, the data must be locked between the two to avoid intermediate change by another process; this may not be possible in the physical environment.

IDENTIFY COMMON PROCESSING

Common processes are identified early in the project. In Stage 3, several common processing elements will have been identified and carried forward. Examples from SS plc would be processes such as accessing a Delegate Bookings record. Several functions may need to perform this task, to create it, modify, or delete it. Each of those functions will show that processing in the Stage 3 models.

Common processes at function level or event level are not carried forward. By the end of Stage 6, the only common processes to be implemented are at the level of database processes, I/O processes, or particular algorithms that can be invoked by many processes.

Superfunctions and functions

Systems can be represented as a network of functions within superfunctions, database processes across functions, and I/O processes across functions. Database processes may well reference common processes, as may the I/O processes. By 'superfunction' I mean a set of related functions joined in a dialogue. In the following example from SS plc (see Fig. 19.3), the superfunction name is Bookings Office, the User Role that governs a particular set of dialogues. This is an example, not a rule. Another candidate superfunction in this context might have been named Delegate Bookings, instead, and handled all the functions to do with accepting, confirming, cancelling, or rescheduling Delegate Bookings. In such a case, the input to a superfunction should contain the data items common to all its subordinate functions as well as the specific input data items for the trigger or event.

So far I have discussed the FCIM in terms of on-line functions. Off-line functions too need to be classified and included on the FCIM. They may be standalone functions, or several may be combined into a superfunction by being batched together in a periodic operating schedule.

Functions and Logical Database Processes

There is a many-to-many relationship between functions and events. Each function may be triggered by several events, while each event occurrence may be responsible for several functions. This leads to Logical Database Processes being regarded as common components across functions.

Logical Database Processes and common processes

Common processing elements inside the Logical Database Processes should at Stage 6 be extracted and defined as a subroutine. This includes invocations to the PDI from different physical processes.

I/O processes and common processes

Some routines that occur commonly are peculiar to I/O processing, such as altering

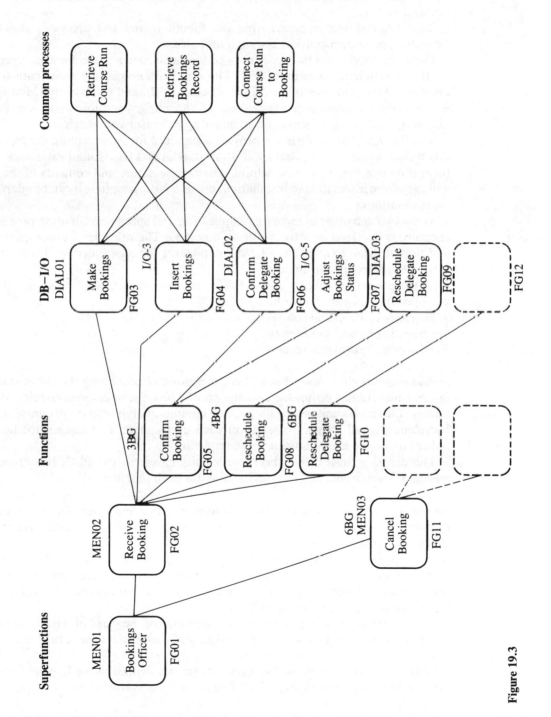

Superfunctions **Functions** **DB–I/O** **Common processes**

Figure 19.3

311

the format of input dates, or carrying out calculations on input data, or for output reports.

Such routines that appear across the functions and I/O processes should be extracted and implemented as common modules.

Figure 19.3 illustrates this network of components, down to the common processes for the superfunction Bookings Office. This representation is only a schematic to help understand the principles of the FCIM, of course. The end result of this Module is a set of coding specifications, regardless of the implementation tool, and such a diagram is not going to help a programmer or generator very much.

The *SSADM Manual* does not provide a standard for documenting the FCIM, as this would depend on the local application standards. I have found variations of the form shown in Fig. 19.4 to be helpful. The precise nature and contents of the form will vary from installation to installation, but as a basic template, it can be adapted to most situations.

These two activities, of removing duplication and specifying common processing, encompass one level of activity during Step 630. The other level encompasses the addition of detail to all of the FCIM components for each function. There are five sets of activities to perform here:
1.   Define success units.
2.   Specify syntax error handling.
3.   Specify controls and control errors.
4.   Specify physical I/O formats.
5.   Specify physical dialogues.

Define success units   A success unit is the portion of processing that must stand or fail as a unit. It may be just the update part of a process, or just one enquiry. For an update, the success unit incorporates the presentation of data to the program and the completion of all changes of state in each entity affected. Changes to the state indicators on the ELHs indicate this completion.

The designer must define the success units based on the PDS, the Processing System Classification, and the specification for each function.

Specify syntax error handling   The definition of syntax error handling is not carried out until Stage 6 so that the constraints and the facilities of the physical environment can be included in the definition.

Syntax error handling is one of the opportunities for specifying reusable code: if an error is found, a code should be used to retrieve the error message from the database, rather than to build the message into each and every program. This reduces the maintenance overhead considerably.

The nature of the syntax error is the definition of the class of data item format (alpha/numeric, etc.) and range (e.g. valid range for numeric must be between, say, 16 and 65).

Semantic errors, or database integrity errors, are defined in the Logical Database Process Design activity in Stage 5, and so are not our concern here.

Specify controls and control errors   'Control' covers two areas in this context: navigational control and control of data errors. In both cases, the classification and

Function Component Implementation Map SS plc

Fragment Ref. No.	Description	Class	Cross-reference	Uses (Frag. ref.)	Used by (Frag. Ref.)	Implemen-tation	Program Spec.	Success unit
FG01	Bookings Office	S	MEN01	FG02, FG11	–	4GL	–	–
FG02	Receive Booking	F	MEN02	FG03, FG04, FG05 FG08, FG11	FG01	C	SDM	At update level
FG03	Make Booking	I/O	DIAL01	–	FG02	C	SDM	Completion of fragment
FG04	Create Booking	DB	I/O-3 Booking Received	–	FG02	SQL	–	Update level
FG05	Confirm a Booking	F	Function 3BG	FG06, FG07	FG02	SQL.	–	Dialogue/ update level
FG06	Confirm a Delegate Booking	I/O	DIAL02	–	FG05	SQL	–	Completion of fragment
FG07	Adjust Provisional Booking to Confirmed	DB	Confirmation Received	–	FG05	C	SDM	Update level
FG08	Reschedule Booking	F	Function 4BG	FG09	FG02	SQL	–	Dialogue/ update level
FG09	Reschedule Delegate Booking	I/O	DIAL03	–	FG08	C	SDM	Completion of fragment
FG10	Reschedule Delegate Booking	F	Function 5BG	FG12	FG02	SQL	–	Dialogue/ update level
FG11	Receive Cancellation Notice	F	MEN03 Function 6BG	FG13, FG14, FG15	FG01	C	SDM	Dialogue/ update level

Class: S   Superfunction
       F   Function
       DB   Database update
       I/O   Input/output

Figure 19.4

the Physical Design Strategy should give the designer help in deciding how the errors should be handled.

Specify physical I/O formats   I/O formats are first specified during Stage 3. These are simply the I/Os for Function Definition, without consideration or error-handling. Few of these early definitions last through to Stage 6.

The I/Os relate either to dialogues and reports, or to input and output files.

In Stage 6, I/Os are designed not just for the logical I/O processing, but also for the physical error-handling and the production of reports. The logical I/O Structure will be superseded by the physical I/O Definition; the physical definition includes various constraints imposed by the devices that are used. Such devices include printers, file transfer systems, terminals, and backing storage devices.

The design of the I/O considers not only the data items, but also any ergonomic considerations that local standards stipulate. The Applications Style Guide provides information for the designer. If external systems or devices are used, such as electronic data interchange (EDI) or bank automated clearing system (BACS), especial care must be taken to ensure conformance with these.

Specify physical dialogues   The designer must take the Logical Dialogue Design as a starting point, and expand it using the knowledge of the physical environment. This expansion is to incorporate such facilities as:

1. *Physical Data Groups in Screen Design*   In Dialogue Design, a physical data group is a group of data items that is displayed on the screen; it describes either the entire screen or a defined block within the screen. It may be that a physical data group matches logical data groups, one to one; this is an economic and obvious path to follow if the physical environment supports this, and if no extra update processing is entailed.

    There are four scenarios that may argue against mapping logical data groups directly onto physical data groups:
    (a)   A logical group may be too big for one physical data group.
    (b)   The physical environment may constrain the contents of a physical data group. In this case, obviously if a logical data group represents a greater size than can be handled, there cannot be a one-to-one mapping.
    (c)   The physical environment constrains the processing of a physical data group. Accesses specified on the logical model may not be supported in the physical design.
    (d)   The logical group can be used for several distinct input messages.

    There are two views of data in a dialogue: the input and the display. These are not necessarily compatible. The input data will relate to an event, or an enquiry trigger. Each event can affect many entities, as each entity may be affected by many events. At dialogue level, a screen may display all of the data groups that may be affected by a choice of events, while the input will be a subset of the data items that constitutes a trigger.

2. *Physical Dialogue Design*
    As described earlier, there are three possible strategies for dialogue handling: one event input and processed at a time; groups of events input and each one processed individually; and groups of events input and processed together.

The Physical Design Strategy should enable the designer to choose which Dialogue Design strategy to follow. A great deal depends, as always, on the physical environment. If, for example, the data on the screen can be redefined as a number of physical data groups, the first strategy can be employed. If, on the other hand, the environment constrains the designer to specify a group containing more than one trigger or event, there is a decision set to follow:

(a) If the one data group is given several views—Strategy 1.

(b) If the data group can trigger several success units—Strategy 2.

(c) If the data group can trigger only one success unit—Strategy 3.

As always, the decisions must be taken only in consultation with the User.

3. *Error-handling components* As well as enquiry and update dialogues, components to handle the three main classes of error must be defined:

(a) Syntax errors in inputs

(b) Control errors in input

(c) Database integrity errors

Where possible, these should be generated by the environment. Where this is not possible, the logical dialogues must be extended to cater for them.

4. *Search/enquiry components* The effort required in the logical Dialogue Design to handle searches through data sets and groups depends on the abilities of the physical environment. If the environment is sufficiently powerful and versatile, little effort is required in converting the logical to physical Dialogue Designs.

5. *Dialogue navigation* The designer must specify the navigation paths within each dialogue and between dialogues. Again, the physical environment may be of help. There are standard navigation paths:

(a) Move to the next/previous data item.

(b) Move forward from within one group to the next.

(c) Move back from within one group to a previous group.

(d) Move up from dialogue to menu, or to a higher level dialogue.

Logical Dialogue Design produces a set of Dialogue Control Tables. The physical environment should be able to implement these. If not, the designer must find a way, perhaps using procedural code, or perhaps by defining a physical table for the screen or for each data group.

6. *Linking dialogues–superfunctions* Menus and direct command links allow the combining of dialogues into superfunctions Menus allow a hierarchical linking. Related dialogues can be placed beside each other on a bottom-level menu, so that crossing from one to the next poses no difficulty. It is up to the User to specify to the designer the optimum requirement for navigation and linking dialogues.

Complete Function Specification

The Function Specifications are taken from the Logical Database Process Design. Procedural and non-procedural programs must be specified. Purely non-procedural programs can actually be generated at this point. Generally, it is found that even non-procedural programs need some procedural code inserted, either through a 3GL such as COBOL or C, or through some special-purpose language.

This step must be executed by a designer. Business analyst or systems analyst skills are not the most appropriate for this technical exercise. Some of the Jackson structure clash resolutions will be the province of programmers.

PROCEDURAL SPECIFICATION

If a function is specified procedurally the processes must be specified in greater detail. The main problem is defining how the logical processes can be implemented as physical Modules, and then combined.

Options that can be considered are:

- Separate logical processes in a function
- Input/output subsystems
- Combine processes

These are described below.

Separate logical processes in a function

The Universal Function Model in Fig. 19.5 represents a generic view of the component processes and linking data streams in a function. Functions which are specified procedurally may need this model to be adapted to reflect a complex lattice of fragments. If there is a mismatch between the logical and physical views of the data, again, a Specific Function Model helps to clarify the views. A typical case

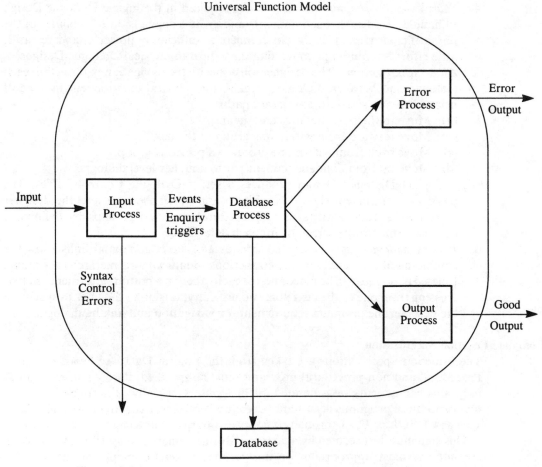

Universal Function Model

Figure 19.5

would be the identification of a structure clash during Enquiry Process Modelling: an interleaving clash or an ordering clash can be resolved using an extra sort routine, to be specified on Specific Function Models for that function.

If a Specific Function Model is compiled it should be recorded with the Function Definition, and may also be incorporated into the Program Specification.

In a procedural specification, the design is carried out using the Jackson design technique of defining input and output data structures and merging them. Chapter 17 describes the possible structure clashes and how they can be resolved.

Input/output subsystems

There may be an operational or performance requirement to store inputs or outputs in a separate database, independent of the main database. This storage of the I/Os would be temporary. The inputs, processing, and outputs would not take place in one success unit, and so it is not strictly a part of the one function.

Where this happens, it is advisable to create a small subsystem alongside the main system, to deal with the inputs and outputs. I/O Structures would define the interfaces between these systems.

Combine processes

Processes can be combined into one success unit or one complete program. They communicate by means of intermediate files. In this way, the Specific Function Model becomes a system run-chart. The use of intermediate files implies that the function specified is an off-line function.

Separate processes, or 'routines', can be combined in one of two ways: co-routines, or subroutine and host. At present, few physical environments support co-routine, or parallel, processing, so the host/subroutine path is the more likely.

Generally, the dialogue process is the host, and the database process the subroutine; this is because the database process can be executed in one uninterrupted pass as a success unit, while the dialogue process will be interrupted by I/O processing and database processing.

However the routines and processes are combined, the designer must return to the earlier specification of success units to be sure that they do not need to be changed. If they do, then of course change them.

An important feature in designing subroutines is the data linkage. Many environments offer the facility to make the linkage explicit. If this is the case, then take advantage of it.

Consolidate Process Data Interface

If the physical environment allows, the PDI should be specified non-procedurally. Most DBMSs allow the defining of logical views of the physical database. The view consists of:

- The operation desired upon the logical data model.
- The operations necessary upon the physical database.
- Non-procedural specification of the physical access path.

If there is no such facility, then the PDI must be implemented as a set of procedural modules.

Access requirements for fragments are compared with the Physical Data Design, after optimization, and any mismatches identified.

The physical keys of master and details for each mismatch are identified, and the sequence of physical accesses determined.

The PDI elements must be fully documented in the FCIM to show how mismatches between the two views are identified and resolved.

If low-level routines are needed, for performance reasons, these must be identified and specified here. If, even then, required service levels cannot be met, then the User must agree to whatever compromise on the targets is reached, and the Requirements Catalogue should be amended to show that decision.

Assemble Physical Design

In this step, all that remains during the SSADM part of the project is to gather all the Physical Design products for validation, for checking of consistency, and for final delivery. When the Physical Specification has been handed to Project Management, and has been formally signed off, the next task is to begin building the database and coding the programs. That is for another team, though. Our formal involvement in the IS project is completed.

Summary

Stage 6 of SSADM can offer only generic help in specifying a Physical Design, because of significant differences between the possible environments. The CCTA subject guide on 3GL interfaces with SSADM gives more clarification on how to cope with such elements as structure clashes in the Logical Design, and how the procedurally defined processes can be translated into code, whichever language is used.

Database design is helped by the SSADM preparatory work on universal first-cut design, and on the classification of physical environments carried out once the environment is known. The suppliers of the physical environments should help in this classification work; if they do not, the designer must find out the storage and performance features directly, instead.

The optimum situation is to find a one-to-one mapping of the Logical Data Design onto the Physical Data Design. Where this does not happen, the PDI must be specified to show the accesses required. This aids the understanding of the development, and eases the maintenance load, by requiring any change to be made to the PDI rather than to the physical storage design.

PART THREE SSADM in use

20. Applications for SSADM

20.1 Aims of chapter

This chapter discusses the applications and situations that are appropriate for SSADM. The topics covered are:
- Suitable applications
- Project procedures
- IT support for SSADM
- Migration from version 3 to version 4

Part 1 described the structure of core SSADM, and its place in the project life cycle. We saw then that SSADM is not the sole activity on the Full Study, and that it cannot take place without considerable management overhead.

In this chapter, I shall look more closely at the overheads incurred in applying SSADM, and how they interlace with the method to produce a product that is as close to the User's requirements and of as high a quality as it is reasonable to expect of any method.

20.2 Suitable cases for SSADM

Notwithstanding SSADM's flexibility and applicability to a wide range of projects, it is not a universal cure-all, to be applied regardless of application area. During either the Tactical Study or the Feasibility Study, the project team must address the question of which method is most suitable for the particular problem, SSADM or, for example, Yourdon, MASCOT, or IE.

There are some criteria to examine how well suited an application is to SSADM. They fall under the following broad headings: data, procedures, culture, project size, management commitment, and technology.

Data

As SSADM activities are centred on a stable data model, it is important that the information central to the application area can be modelled in a Logical Data Structure. If it is entirely free-form text, say, or statistics, it may not be possible to build a stable model. In that case, SSADM may not be the most appropriate approach. If a subset of the information is in one or other of these forms, but the bulk of the information is held in a more structured form, then SSADM combined with a package approach is a perfectly good strategy to follow.

If the target system is a knowledge base for an expert system (ES), it is worth thinking about how SSADM can help. Knowledge elicitation is more complex and heuristic in principle than the conventional fact-finding exercise, and therefore harder for the Project Management team to estimate. However, once the knowledge base is being designed, it is less of a departure from conventional file/database design than it appears at first sight.

SSADM needs to be extended to model the inference processes and storage of rules and knowledge, but none the less, extension is not the same as unlearning in order to adapt. There are a few products and methods for designing an ES on the market now, but none has yet achieved a significant market dominance; CCTA has funded and guided the GEMINI project to develop a method for designing an ES. In time, it is expected that GEMINI will become an open method, as SSADM has become.

Structured Techniques for the Analysis and Generation of Expert Systems (STAGES), developed by Ernst and Young, was designed to be compatible with SSADM. It is beyond the scope of this book to describe how ES methods are designed, and to go into detailed description of the architecture of knowledge-based systems (KBSs). Interested readers are directed to the literature on expert systems, and to Hares (1992) for further explanation. The purpose of this section is to point out that the structure of a KBS does not preclude the use of SSADM.

Procedures

SSADM expects a degree of formality in procedures, to be able to model trans-formations of data, and to identify different functions and user roles. If investigation reveals that no such formality is present and, more significantly, is not required in the future, SSADM may not be appropriate, again. The ease with which top-level Data Flow Diagrams can be drawn will indicate whether or not this is so.

If clear, functional decomposition cannot be carried out, then another approach, such as Checkland's Soft Systems Method (SSM) Checkland (1991), or Mumford's ETHICS would suit the case better. After such a study, there may be a case for SSADM to work on identified formal areas of procedure.

In many sites, the PID has been distinctly unclear, and a preliminary study on soft principles has resulted in high-quality systems that pleased the User community better than an immediate structured analysis that could not uncover hidden organizational problems. After such a soft study, SSADM is better able to address many of the procedural issues that arise as a result.

Culture

To succeed, SSADM must involve the Users for the duration of the project. The analysts and designers will work in close consultation with the User Requirements from the first investigation, through the whole requirements and functions analysis, through to the confirmation or amendment of performance objectives.

Some organizations do not encourage such close interaction between business Users and the IS practitioners. Very often the only contacts occur during fact-finding, to approve options and to sign off the User tests. If the target organization has such a culture, SSADM may not succeed. In such an environment, it is hard to imagine any method guaranteeing any measure of success!

The culture must be one of complete cooperation between the different communities. On the one hand, the IS community recognises the business User's ownership of the system, and so works hard to understand and meet their requirements; on the other hand, the business Users recognize their own responsibility towards achieving the IS system that they require. Thus they do not leave everything to the IS department, only to complain of the shortcomings of the system when it is finally delivered.

I was recently called in to assist on a project where a deadlock had arisen between development staff and Users. My initial, phoned, brief was to the effect that, 'We have a problem with our Users!' This is bad news for any major project, and this particular one was an important component in a suite due for implementation within a few months—or else.

The problem was not with the Users, it was with the project team, who, after initial interviews for requirements gathering, had not approached the Users until it was time to review the Selected Business System Option. I asked them to repeat this, sure that I had misheard, but no, the team had, unilaterally, selected a BSO, without even showing the other options to the Users, and were ready to move on to Requirements Specification. There were political and time-constrained pressures that prompted this, but it was still indefensible. The only people who select a BSO are the Users, after being presented with the cases for and against any particular one on offer.

It was no wonder that a 'problem' had broken out between them. User relations did improve further along the project when I persuaded the team that they must work closely with the Users; it was not even as though the Users were reluctant—they were asking for meetings and consultations. Prototyping helped to restore confidence, by letting the Users see and feel the new system, but the whole incident confirmed what, for me, has always been a key tenet: without User involvement throughout, a development project will never be more than half successful. Even if the technical solution works completely, it will be an imposed solution, not one oriented to the human and organizational needs of the customer. If this culture of participation is missing, therefore, so will the optimum development be missing.

Project size

The cost of using SSADM techniques, with all of the Project Management and Project Procedures overheads, must be justified by the measurable benefits of choosing that path.

If the system is small (a subjective judgement, perhaps), the cost of employing the rigour and complexity of SSADM may outweigh the benefits of getting a simple system up and working fast using a simpler (or no) method.

If the system is medium to large, with a degree of complexity, the use of SSADM will pay for itself in ensuring a product delivered to a planned timescale, and to an acceptable level of quality. As a rule of thumb, the greater the level of complexity, the longer and more risky the test period will be. Without a rigorous front end like SSADM, this test period may be protracted, covering as many paths as possible, against pressure from managers to deliver as soon as possible.

Using SSADM, the specification should be so precise that even complex systems should find testing a matter of just trying code, rather than finding holes in the specification. That such holes do appear was shown dramatically in early 1993 when TAURUS, the proposed Stock Exchange share transfer system was cancelled abruptly, amid recriminations and resignations.

There were all sorts of reasons for the TAURUS failure, but a key one was to do with how SSADM was 'used' by Project Management. In fact, although the project team were nominally applying SSADM, it was in name only; this was shown as much as anything by the fact that operational testing had begun even before the specification was complete. A reason for this was the sheer size and complexity of the project, which, without rigorous Project Management, made coordinating activities and stages riskier. There were other factors involved in the debacle, including ever-increasing security demands insisted on by the junior minister with special interest for the project, as well as scattered and sporadic project coordination (*Financial Times*, 12 March 1993). TAURUS proved a case study warning of both poor analysis practice and poor Project Management practice.

Version 3 was modified with a fastpath, micro-SSADM, which was for small systems, but only if designed by a very experienced team. At the time of writing, there is talk among some suppliers and User Group members of introducing a fastpath for version 4, and indeed, different versions for different sized projects. This is at once a laudable aim, and a very dangerous path to follow, particularly when project managers and steering committees are under pressure to deliver to impossible timescales in order to preserve margins.

SSADM is flexible enough to be tailored to suit local objectives and targets, but any move towards a fastpath should be attempted only by experienced and knowledgeable analysts. The cross-checking of techniques, and their interdependence, can be seriously compromised by arbitrary management decisions to leave out this technique or that.

Management commitment

An unfortunate misconception of earlier versions of SSADM was that it was a Project Management method. This was never intended to be the case; CCTA have never propounded that view of the method, neither has any training organization. However, many managers in the industry appear to have been of the opinion that if a couple of members of a project team have attended an SSADM course, the whole team can carry out an analysis and design project without any interaction with formal Project Management. Whenever loud condemnations have been made of the method, they have usually come from just such a manager.

As I stated in Chapter 2, SSADM occupies one part of a project life cycle. The version of SSADM described here makes it harder to allow such misconceptions. It is not a quick and simple way of installing a system, an algorithm that can be applied by the blind and inexperienced, or the stupid. The practitioner must be trained and experienced, and have full support. Whereas this should always have been the case, I have seen examples of projects that have been resourced and planned quite wrongly.

Version 4 of SSADM has made explicit what should be common practice. The information highway and the interfaces with Project Procedures mean that an organization which uses SSADM has to commit resources to it, and plan its progress. The practitioner team is no longer the only set of actors taking part in the enterprise.

This all presumes a serious commitment to the success of an SSADM project, on the part of management. If they are not prepared to provide the full level of support, and ensure that all required Project Procedures are in place, and the staff all have full training, then perhaps SSADM is not going to suit the culture of that organization.

The odds against any systems development being successfully undertaken in such an environment are high, regardless of what method is used, be it SSADM or SoP (seat-of-the-pants).

Technology

SSADM's purpose is primarily to design large, databased information systems, processed on a single site. This was the prevailing DP environment when the method was first developed, and has been the most common environmental demand behind each update, including the current version 4. By the time that the invitation to tender was produced for version 4, however, other technologies and other approaches to IS requirements were growing. This does not mean that SSADM cannot be used to design such systems, but thought must be given to how it can best be made to serve the different requirements of such technologies.

Developing technologies, such as object-orientation, can be developed using SSADM, but the method must be tailored to the needs of the technology. Similarly, small systems and distributed systems are not specifically catered for in the structure and procedures of SSADM, but none the less can be developed, with intelligent tailoring of the method. SSADM is flexible enough to cope with particular approaches and strategies of IS development. CCTA publications are available to describe how best to adapt SSADM to small systems, distributed systems, and the growing move towards object-orientation.

While small systems might present a management problem, because of disproportionate over heads, technologies like distribution pose a more philosophical problem to structured methods. A major problem with applying SSADM to distribution, for example, is in the question of replication of data across separate sites and ensuring its integrity. In a single-site system, there is no such problem; in a distributed system, there is the problem of entity occurrences in satellite sites existing in a different state to the same occurrence held on the host database. This is not a problem peculiar to SSADM; it belongs to distributed systems, however they are designed. SSADM's problem is how to model this using its standard techniques, particularly ELHs. Standard ELH techniques would not reflect this dual state, so a new way of modelling this must be found, based on location of data.

20.3   Circumstances where SSADM may be used

This is at first sight a curious heading; after all, SSADM is a method for analysis and design, so surely that says it all? As it happens, no. In general terms, there are three situations where a practitioner team will be invited/instructed to use SSADM, either in full or in part:

1.   In-house development of information systems.
2.   Bidding for a contract to develop a system using SSADM.
3.   Third-party development work for remote clients.

In-house development

In-house development teams are likely to use the method according to the manual to develop their own company's system, with the User constantly on hand, as I have indicated they should. Mistakes that the team make in early projects will be useful learning points, and the culture that supports SSADM, including good Project Management practice, will grow. The practitioners frequently follow a project through from start to finish, and so gain expertise in all parts of the method, rather than find themselves specializing in just Requirements Analysis or Logical Design.

Local application standards will be familiar to the teams, and newcomers should find support during their own learning curve. Doubtless there will be time pressures that cause discussion on fastpaths, or short-cuts, but as all parties should be working towards the same corporate success, there will quickly grow a strong pool of experience that can be shared between projects.

Bidding for contracts

Since the mid 'eighties more and more systems have been developed not by in-house teams, but by third-party suppliers, who may pick up the project at different parts in its life. There are at least three contract points in SSADM where new invitations to tender (ITT) can be published: after Feasibility; after selection of Business System Options; and at the end of Requirements Specification. At any of these points, a fresh team can take over the project, and will rely on the successful use of SSADM by its predecessor.

Thus, when a third-party supplier does make a bid for the project, it often finds that it is working on the back of an earlier study that may or may not be well carried out. Frequently, especially on government contracts, the ITT will specify some SSADM involvement, and ask for sample diagrams to be submitted as part of the bid. This creates a problem if the earlier study, carried out by anonymous and unobtainable teams, contains errors or is incomplete. This is not an unlikely or even an uncommon situation. In such cases, the team preparing the bid switch between amused dismissal of the earlier team, and concern that the customer is using this as a test of the bidder's real SSADM knowledge and skill!

I know from my own experience of an ITT to carry out the design phase of a sensitive government contract. As part of the bid, I was required to produce a number of Update Process Models (UPMs), based on an LDM and a series of ELHs. I was not permitted to contact the Users directly, and the deadline was just two weeks away. The team who carried out the analysis had either not enclosed all the ELHs and ECDs, or not completed them. As a result, I had to spend much scarce time trying to create ELHs and ECDs, with little clear indication of the business rules to model, just in order to create the inputs to the documents I had been asked to produce.

The local project manager, who had wide and good experience in managing software projects, but no experience of SSADM, was designated as go-between for me and the User, 300 miles away.

This was not a good situation, and there was a strong case for deciding the risk was too high. Political reasons dictated that we should produce a bid, complete with UPMs and Timing Estimates.

While the other aspects of the bid were being prepared by other team members whom I did not meet, and was not in a position to liaise with, I worked on the rogue ELHs, ECDs, and UPMs. As it happened, the original team had produced their own Event Catalogue, with cross-reference to Function Definitions. This gave me an insight into the business rules that were being modelled, but, as with the rest of the material, it was incomplete, and needed the project manager to phone through for clarification. This all took time, which was already very tight; it also worried the project manager, because he did not feel competent to understand my questions, or the Users' answers!

I completed the tasks, and the bid was submitted on time; all the good practices of SSADM were set aside for expediency. This, as I say, was an invidious position to be put in, but in the circumstances was unavoidable. In a Full Study, this is unlikely to happen, but in a bid situation, it can be all too common. A major weakness was in appointing a project manager who, however good he was in other regions (and he was extremely good), simply did not know or understand SSADM techniques. This made for difficulties in estimating and in clarifying requirements.

If a practitioner is faced with this situation, only experience can help. Common sense and self-preservation may dictate a refusal to take part, but in the 'real' world, there are other pressures that compel seemingly impossible tasks to be undertaken. It is the Project Board and project manager's decision as to whether these are done or not. The practitioner must use intelligence and understanding in carrying out the tasks. In these cases, following the manual to the letter would probably be a mistake; the practitioner must draw on SSADM's flexibility to achieve the task, and must also identify the risks involved in the approach, so that the project manager can relay these to the clients.

Third-party development

When the contract is awarded to a successful bidder, there are other potential risks involved. As with the bid situation, the previous documentation may be incomplete, or inaccurate. Another common problem is that with the change of project team, there may come a change in requirements, which would affect all the work that had gone on before.

At the moment of starting work, the practitioner may discover that there is no house standard for presentation of work or quality review procedures. If so, the contractor must have some standards already prepared, so that no time is taken up negotiating them. If house standards are present, they should be adopted immediately.

A major problem in working in the third-party environment for a long time is that contract staff rarely see a project through from beginning to end. Thus, they may gain expertise in a fairly narrow domain, say logical data analysis, but lack detailed experience in other aspects, such as converting that logical design to a physical database. None the less, as with any SSADM environment, the depth and type of experience is often less important than ability at analysis and interpersonal skills.

20.4   Project Procedures

In Chapter 2 I described the two parallel streams of activity that take place during an IS project. SSADM represents one of these streams, but does not itself cover the entire life-cycle, or all of the activities. This section draws attention to some of these ancillary activities that interface with SSADM.

I shall not try to give a full description of any of these activities, or Project Procedures, in this section. The CCTA publications cover them in the required depth. I shall limit myself to enumerating the principal procedures and describing their place in the method.

Project Management

The current structure of SSADM is designed to give optimum support to Project Management, in that it interfaces with the management using predefined products. The assumption is that the chosen Project Management method or structure will be similar to the government standard, PRINCE. While PRINCE is a specific product, its structure and philosophy can be applied generically. Like SSADM, it is oriented towards products, rather than monitoring activities.

Broadly, a method like PRINCE employs a project board, comprising an executive, a senior user and a senior technical representative. Between them, these three roles represent the key interests in the project: management, the User community, and the IT community. The board's function is to control the whole project. They will be in a position to apportion resources from their own areas, as required.

The board will appoint a project manager (PM), who will oversee the day-to-day running of the project. The PM is described as the information highway in the SSADM structure model, acting as the interface between the practitioner team and information and control, or the project board.

Below the PM will be a separate manager for each activity, whether module or stage. That activity manager's responsibility is to ensure that targets are met, and that products are produced to the required quality and within budget.

The activity manager will be supported by the activity team responsible for the production of the deliverables of that activity.

There must also be a project support office to provide whatever support services are required, and also perhaps to assist with the resourcing of the project.

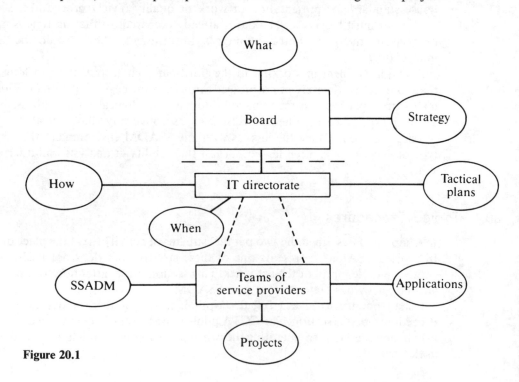

Figure 20.1

Table 20.1

Control point	Who exercises control	Triggering event
Project initiation	Project board	Authorization for project by project sponsor(s).
End module assessment	Project board	The end of a stage in the project technical plan.
Mid-module assessment (unplanned)	Project board	An explanation plan has been prepared; or the next stage needs an early start.
Project closure	Project board	All products have been delivered.
Quality reviews	Review chairman	A product has been completed.
Mid-module assessment (planned)	Project board	Arrival at planned point.
Checkpoint meetings	Module manager	Arrival at planned point.
Preparation of highlight reports	Project manager	Arrival at planned point.

Other roles are suggested within the Project Management framework, but I will not go into that issue further than to say that the above structure, at least, is necessary for an SSADM project. Local standards and culture will dictate the rest of the framework.

Figure 20.1 gives a view of such a management structure.

PLANNING AND CONTROL

The levels of management overseeing the project, either in its entirety, or at activity level, must draw up plans which will normally be in two parts: technical plans, showing the activities and dependencies; and resource plans, itemizing and costing the resources required for each activity or phase.

Among other considerations, this is the time that the quality of each product is determined, so that it can be built into the plan. Without determining the quality criteria, monitoring the project will be less effective.

I shall not go into further detail of how planning is to be carried out, or approved: that is outside the scope of this book, describing as it does a separate activity from systems analysis and design.

Control of the plan once the project is in progress is another key function of Project Management. It is suggested that the control points in Table 20.1 provide the most useful controls for project assurance. The table is taken from the *SSADM Version 4 Reference Manual*. Again, I shall not describe the different procedures and issues for the controls further. The CCTA publication on Project Management describes and discusses the issues in sufficient depth for the practitioner.

Estimating

Whether the environment is in-house, third-party contract, or a bid, the project board and customer need to know how long the project will take to complete, and how much it is likely to cost. Estimating has always been problematic in systems analysis and design, because of the variable factors, such as experience of resources, availability of resources, level of staff turnover, and so on.

SSADM recognizes the difficulties in accurate estimating, and by breaking down the project into chunks of modules, stages, and steps, and further, by identifying products from each of these bottom-level chunks, makes it easier to estimate the efforts required to produce the individual deliverables.

CCTA have produced a publication to give guidance in this area as well, which describes the rules of thumb that can be employed. One of the most valuable aids to estimating, though, remains a manager's experience and knowledge of the project team.

Such experience should be held in a quantifiable form on a database. Easy to say, but in an industry which contains a high proportion of nomads and contractors, hard to do at all reliably!

Software estimating techniques have been explored and developed for many years. Barry Boehm's constructive cost model (COCOMO) is a much-used approach to the problem of estimating. COCOMO describes the project costs at three levels. First *basic COCOMO*, which gives a formula for estimating the project cost in terms of delivered lines of code (LOC). By itself, this is not very helpful, ignoring the code generation platforms and skill levels of practitioners. Secondly, *intermediate COCOMO* addresses these inadequacies and brings in staff experience levels, constraints of hardware, and the use of particular variations in program development practice. Finally, *detailed COCOMO* recognizes that different phases of the project entail different degrees of complexity and skill. Sets of formulae using these various cost drivers produce estimates that are, in 68 per cent of cases, within 20 per cent of the actual costs incurred. PMs and auditors will recognize that this is no mean achievement!

Another approach to estimating that has been used on SSADM projects is Function Point Analysis Mk 2. Again, this is a detailed mathematical approach, with a complex series of formulae based on weighting factors.

Whichever method of estimating used, it is the PM's responsibility (for which, many practitioners heave sighs of relief!), and so the PM must have details of as many of the variable factors as possible. CCTA provide guidelines on factors such as complexity of DFDs, leading to numbers of functions, events and so on. The less algorithmic factors, such as breadth and depth of experience of team members, familiarity with the particular environment or with similar systems elsewhere, political pressures and constraints, level of non-functional requirements such as security requirements and so on, are down to the PM to evaluate. A good database and/or spreadsheet is necessary to build up a picture of these factors, and to allow reasonable estimates to be made from them.

Many otherwise good project teams and SSADM teams have experienced problems simply through the lack of good estimating techniques or practice available at their installations.

Configuration Management

Configuration Management is an important part of controlling a project. The more complex the project, the more essential is good Configuration Management (CM).

By CM, I mean a set of procedures for controlling all of the products from a project, ensuring that only current versions are produced and are in place, and that no changes can be made to any product without high-level approval and documentation. This has proved necessary after many unhappy experiences when a change has been made without authorization, and without all the knock-on effects being considered.

There are four sets of procedures involved in CM.

CONFIGURATION IDENTIFICATION

This means identifying a product that must be signed off and controlled. Each item on the SSADM Product Breakdown Structure (Chapter 1) is a configuration item that must be subject to such control.

Products are only placed under configuration control when they have achieved sign-off, having met the predefined quality levels.

CONFIGURATION CONTROL

Once the configuration item is accepted it is subject to both physical control, whereby the physical document is kept in a project library, and logical control, whereby amended versions are labelled and maintained. Related sets of items are kept together as a 'release', ready to be used as input to a subsequent SSADM activity within the project.

CONFIGURATION ITEM CHANGE

After an item has been accepted for configuration control, it may change for any one of several reasons: a new requirement by the User; or an error shown up by testing or by validation against another SSADM product, for example. The ramifications of the change must be investigated, and any other items affected also changed accordingly. The change request must be sanctioned at high level, not by an ordinary team-member. As well as the technical ramifications, the project plan must also be investigated for resultant delays or changes to the critical path.

CONFIGURATION AUDIT

Audits will be held, perhaps periodically, certainly before major decision points, to ensure that the actual products in the library tally with Project Management information on them.

Quality Control

Quality is an inherent part of SSADM, yet it does not have a fixed place in core SSADM. It is very much a feature—and function—of Project Management.

The exercise of quality control is aided by the establishment of a set of quality criteria for every product delivered by SSADM. There are two basic forms of Quality Control: formal and informal. Informal checks have nothing to do with management, and are not sufficient to achieve sign-off. They are valuable, however, in helping to ensure that formal quality assurance (QA) reviews are productive, and result in acceptance of the product.

Formal QA reviews follow a pattern, with actors and procedures that should be standard. The actors are:

- *Presenter* Usually the person who has produced whatever is being reviewed. The presenter will introduce the product, its scope and description, having distributed copies to all participants well beforehand.

- *Chair*   Ensures that the review is conducted fairly and thoroughly. The chair may note all the comments and criticisms of the item under review, or there may be a separate scribe to do that. One main responsibility is to see that the review identifies problems and errors, but does not attempt to solve them. This can be one of the most difficult tasks, given that the composition of the review panel will be problem-solvers and experts in the functional area!
- *Reviewers*   Their remit is to ensure that the items submitted are both accurate and complete, certainly up to the level laid down in the Product Description quality criteria. They will identify errors or weaknesses, and accept ownership of them, thus ensuring that the corrections are made before the item is accepted.

Each installation will have different ways of conducting such reviews and following up the action points, but the structure above is required for a successful review. The Product Breakdown Structure lists quality products as well as technical products. These quality products are all from the formal Quality Control reviews.

Risk Analysis and Management

The risks in question in this set of Project Procedures are twofold: those associated with the security and integrity of a business's data; and those linked with the successful conduct of the project itself.

Just as SSADM is compatible with a Project Management style in the nature of PRINCE, so it is compatible with a Risk Analysis and Management method called CRAMM, the CCTA's own standard. As with Project Management, the standard is not a mandatory one, but rather a generic guide.

CRAMM is one method of Risk Analysis, but most approaches follow broadly similar lines. There are always several elements to Risk Analysis and Risk Management:

- Identifying sources of risk
- Assessing the probability of occurrence
- Assessing the possible impact of the risk
- Developing strategies or responses
- Implementing strategies or responses
- Monitoring and controlling the risks

The hardest part of this is normally identifying the sources of risk. There are a number of areas to examine when completing a Risk Analysis, for example:

- Technical
- Management
- Financial
- Schedule
- Customer

The team will examine each of these sources to identify where a risk may occur. For example, if sound Project Management procedures are not in place, there could be delays in a particular stage in the life-cycle. The likelihood of such a delay should be gauged, and then the impact calculated. A reason for the procedures not being in place might be the relative inexperience of the designated PM. If this is known, then a risk is identified which can be assessed.

Another example could be in the technical platform selected to implement the system. Often a technical solution is selected, or imposed, because of political or contractual reasons, even though the designer recognizes that it will not provide all the service-level requirements the User wants. There is a possible risk identified here, which must be evaluated, and then its impact assessed.

In all such cases, the identified risk will be built in to the estimates, and the project board appraised. The avoidance or management of the consequences of such risks will then be up to the board.

In the SSADM manuals, risk is taken to refer to security or integrity dangers in the complete system. If the system is especially sensitive, such as in a military or intelligence environment, Risk Analysis will tend to concentrate in this area. Even so, other factors which affect the success of the development work should be considered as well, although they would stay at Project Management level rather than be filtered down to the team.

It should be up to the project board as to how they address these risks, but the exercise will have an impact on particular stages of the SSADM study.

The areas where Risk Analysis will be most relevant to SSADM are outlined below.

FEASIBILITY STUDY

While not a SSADM technique itself, Risk Analysis is applied at this stage to help determine the feasibility of a proposed project. Some of the issues involved will revolve around the levels of security required in a viable project and the financial implications of either proceeding with change or deciding not to change. These issues include:
- Project risks to do with progress of the study
- Experience/lack of experience with the team
- Use of subcontractors
- Reliability of prototyping platform
- Reliability of testing platform
- Constraints on timescales and resources

REQUIREMENTS ANALYSIS

One of the focuses of the investigation into the current environment is the levels of security achieved now, and how they are met. The findings are either carried forward to the required system as they stand, or have refinements built in to the specification for the new system.

BUSINESS SYSTEM OPTIONS

All the requirements relating to security should be incorporated into each option description, and be submitted to the board as a part of the supporting documentation.

In the presentation of the options, the analyst should identify and highlight any project-related risks to help the User make a decision as to a particular route to follow. As timescales and resources are likely to be dictated by User requirements and priorities, it is only right that the Users should recognize the risks, and make their selection accordingly.

TECHNICAL SYSTEM OPTIONS

The implications of required security levels are made explicit in TSOs, with descriptions of how operating software, applications software and hardware will be designed or chosen to meet the requirements. As with the BSOs, if the platform does present any risks in capacity, reliability, compatibility, etc., these should be identified and assessed as part of the TSO.

PHYSICAL DESIGN

At testing stage, the security and integrity requirements will be trialled, so the testing strategy must take note of them. Similarly, the operating instructions and User manuals will describe the measures to be taken.

If time constraints or complexity mean that the testing cannot be exhaustive, and some functions will have to be taken almost on trust, the risks involved for each such function must be assessed and reported to the project board for their sanction.

Once all the sources of risks are identified, they should be evaluated, perhaps on a scale of low–medium–high, for both likelihood and impact. A pro forma for such an exercise is shown in Figure 20.2.

Capacity Planning

The planning in question is to incorporate a new application on to existing hardware and software, or alternatively, to investigate the impact of a new computer configuration. The configuration may be standalone, or it may be a part of a network—or even the whole network.

While this activity takes place outside SSADM, it requires three inputs from the core activities:

1. Information on processing of data and transfers of data.
2. Volumetric and frequency details, to help select a hardware configuration.
3. Information on required service levels and non-functional requirements.

WHERE USED WITH SSADM

Step 410—Define Technical System Options   The service-level requirements and proposed volumes are tested against the suggested hardware configuration and DBMS capabilities. These are a part of the Technical Environment Description.

Step 420—Selection of Technical System Options   Only two or three TSOs are put up for selection, out of the half-dozen or so initially proposed. Once the short list is decided, the Capacity Planning techniques are applied with greater rigour, to give the selection board as much detailed information as possible, to assist with the final decision.

Stage 6—Physical Design   At the final stage, the hardware configuration and target DBMS are known, and the Logical Design is converted to map onto them. The requirements must be tested against the final design, on the target configuration. Capacity Planning enables this to be done accurately, so that the design can be tuned to meet performance objectives before the final conversion and loading.

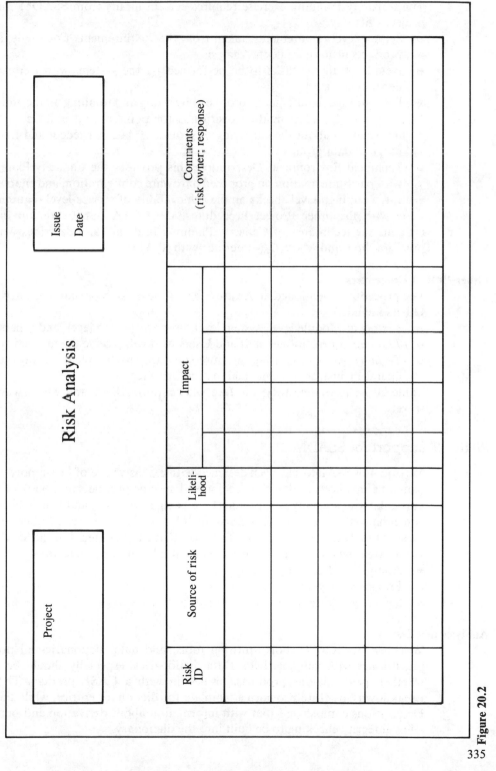

Figure 20.2

'HOOKS' INTO SSADM

The Capacity Planning exercise requires certain inputs from SSADM early in the project's life:

- Service-level requirements, taken from the Requirements Catalogue, and other products of Requirements Analysis.
- Details of all the tasks to be performed by the system, with frequencies and dependencies.
- Data storage requirements, produced by sizing and totalling all the entities in the data model, and estimating overheads for pointers and indexes. In addition, information about the clustering of groups of data is needed, and frequency of usage of data groups.
- Technical Environment Description: this provides the Capacity Planning team with basic information on proposed hardware configuration, and enables them to carry out high-level checks on the achievability of service-level requirements.

As with the other Project Procedures, the CCTA publication on the subject explains the techniques of Capacity Planning in detail, and describes with greater precision how and where they interface with SSADM.

Other Project Procedures

The procedures mentioned above are a few of the most important ancillary activities. Others include:

- *Testing* at Module level, system level, user acceptance level, and others.
- *Training* of practitioner staff and Users who will receive the finished system.
- *Technical authoring*, using in-house standards to complete the various reports, operating instructions, manual instructions etc.

Table 20.2 shows how they interface with core SSADM, and which way the flow travels.

20.5   IT support for SSADM

Version 4 of SSADM has been designed with extensive use of IT support in mind. It would be invidious to name or recommend specific products in a book of this kind, particularly as so many new products are being developed, and will be both released and enhanced during the life of SSADM V4. I shall confine myself, therefore, to discussing briefly the generic forms of support that can be offered to the SSADM team.

The areas where IT support is most useful fall into three categories:

- Analysis and design
- Project Management
- System generation

Analysis and design

Workbenches, CASE tools, drawing tools, and data dictionaries all provide the practitioner with valuable help. Data dictionaries, especially should be available, whether as standalone products, or tied in with a CASE product. The Product Breakdown Structure provides a template for dictionary entries, while the Product Descriptions complement that with information about derivation and purpose that allow integrity checking to be built into the dictionary.

Table 20.2

Procedure	Interface		Level
Project Management method	I	O	Modules, stages, via product/plans/structures
Standards	I		Moldules via Installation Standards
Quality Control	I	O	Modules stages via products/descriptions
Capacity Planning	I	O	Sensibility/BSO/TSO/via option descriptions
Risk Assessment	I	O	Feasibility/BSO/TSO/Physical Design, via option/data/Function Descriptions
Talk-on		O	TSO/Physical Design via TSO and PD products
Physical authority		O	BSO/TSO/Physical Design, via User Requirement/option descriptions
Testing		O	TSO, via TSO/User Requirements
Training		O	Feasibility/BSO/TSO/Physical Design, via option description and design products

At the least, the practitioner will need access to a good drawing tool that supports the notations of SSADM, and possibly other notations that support the local style guide for Project Management. SSADM is very rich in graphics, and to attempt all of the diagrams on paper, and to subject them to frequent revisiting and revision, would be a considerable overhead on a project.

Coupled with the drawing tool should be a good word-processor for the supporting documentation and reports that accompany the diagrams. A good CASE tool will provide both drawing capabilities with word processing and validation checks, so that a Data Flow Diagram, for example, can be validated against the rules for the technique. Ideally, cross-technique checks should be built in. Thus, the Data Store/Entity Cross-reference would be built in, and the checker would ensure that all attributes on the LDM were created by a process on the DFD, and that every entity was created and deleted, at the least. This would also cross-refer to the ELHs, so that all three viewpoints support each other, as they do in SSADM.

Some analysis tasks should be automatic, such as deriving Logical Groupings of Dialogue Elements, or Effect Correspondence Diagrams, or adding state indicators to ELHs. Such tasks should be carried out by the CASE tool used from the data already held in the dictionaries.

Project Management

A number of Project Management tools are available on the market, and while none are specific to an analysis and design method, SSADM's hooks into management make some features especially relevant. The Product Descriptions for an SSADM project, for example, allow Capacity Planning and Risk Analysis and Management to be automated.

The modular structure allows easier Estimating to be carried out. A spreadsheet package could allow estimates to be made, and tailored where necessary, and for this information to be fed into a Project Management package. The spreadsheet would need to be fed from a database based on previous project experience for the team.

The estimating package may be based on COCOMO, or function point analysis, but whichever model the team uses, the package should offer support.

System generation

If a 3GL environment is used for development, the end product from Stage 6 of SSADM is a specification for programs and design for a database. If a 4GL or application generator is chosen as the development path, the end product is a working system. SSADM Logical Design is intended to hook into a 4GL and generate the code, thus saving significant development costs. The actual interface is not defined inside SSADM, as obviously there are as many interfaces as there are products. It is expected that vendors of such products will produce their own guides or interface specifications.

One quasi-problem that has occurred in project work is that many 4GLs are based on non-procedural code, yet the Logical Design is very much procedure oriented. Therefore, runs the argument, UPMs and EPMs should not be drawn if there will be a 4GL implementation. This is less of a problem than some project teams have made of it. The UPMs and EPMs still provide a hook into a generator, even a non-procedural one; more importantly, they verify the data model and requirements. Any technique which concentrates the designer's mind this way is valuable.

The SSADM products that help in such interfacing are the Product Descriptions, supplemented by the Product Breakdown Structure. The entity–relationship–attribute models of SSADM, published by CCTA, provide a base for tool-builders to design an interface for the method.

The structural model and the dictionary have both been designed to enable the teams using SSADM to make use of IT tools through the development cycle. The level of support actually used depends on local policies and management decisions. Basic facilities that should be used are graphics tools and word-processing. Lack of an automated data dictionary, in some form or other, will slow up the progress of the team, leading to longer integrity checks, duplication of information, and increased data entry.

Ideally, a CASE tool covering most of the cycle and made specifically for SSADM version 4 should be used for optimum support.

The University of Central England is the site of the SSADM Research Centre, which tests CASE tools for conformance with SSADM standards. The centre awards a star rating to the various products that are tested; the ratings do not reflect quality, but the range of SSADM products and facilities that are covered. Periodically, the Centre publishes findings and ratings from a tranche of products that have been tested. Any organization looking for IT support for a SSADM product should examine the findings from the Centre as the first part of its investigation.

20.6   Adoption of SSADM

There are two types of organization looking to adopt the current version of SSADM: those new to SSADM, and perhaps to any analysis and design method, and those who already use SSADM version 3 and wish to migrate to version 4.

Those in the first category will carry out their own Feasibility Study on how suitable SSADM is for their purposes, and whether their culture and organization will make the best use of the method.

Those in the second category will look at two issues:

1. How can the new version be introduced, complete with the procedures and software support?
2. Can a project team currently using version 3 migrate mid-project to version 4?

In this section, I shall look at the second question. The first is, again, a management issue.

There are two key areas of difference between version 3 and version 4:

- Architecture
- The move from an activity-based method to a product-based method

There are, of course, many more differences in detail, but these are the two significant broad areas to address when investigating a possible migration.

The modular structure of version 4 provides a number of entry points. Figure 20.3 shows the possible migration paths between the two versions. The end of Feasibility clearly provides a common starting point for Full Study, and the products from V3 can be input to V4 with minimal changes.

The next point where migration can occur is after the Requirements Analysis Module in V4, and after Stage 2 in V3. Not all of the products will be in place to begin Logical Systems Specification, notably the data model enhanced by RDA.

The last possible migration point is between Stage 4, Data design in V3, and Module 4, Logical System Specification in V4. If this path is selected, products from V3, Stages 2, 3 and 4 must be in place. All will be needed. The exception could be the V3 Technical option: the first activity in Module 4 is Technical System Options. All other products from 2 and 4 must have been completed.

One place where migration will not work is from V3, Stage 5 to Module 5, Physical Design. The inputs to V4 Physical Design are very different in content and standard so that the V3 LUPOs and LEPOs (superseded by UPMs and EPMs) would be meaningless to a V4 Physical Process Design.

A problem with migrating, at any stage, is the difference in notation. The notation for LDM is very different across the versions, and so must be amended to meet the V4 standards. This is important with this particular technique, as the same diagram can exist in each version, but with a very different interpretation between them. Entity Life Histories will have been completed in version 3, but not Effect Correspondence Diagrams; ELHs will not have operations added, but may have state indicators included already. Function Definition will not have been carried out for V3, and the Dialogue Design will have been executed using old notations and philosophy. For this path to be successful, a great deal of managed work must be performed on converting V3 documentation to V4, and on Configuration Management.

The lesson from this is that migration is possible, but must be carefully managed. If management have made a decision that version 4 is to be adopted, and have decreed that forthwith a V3 project will be completed using V4, the project board and PMs must insist that resources are made available for carrying out an audit of products delivered from V3, and for converting notation of one technique to another.

The Product Descriptions and Product Breakdown Structure enable the project

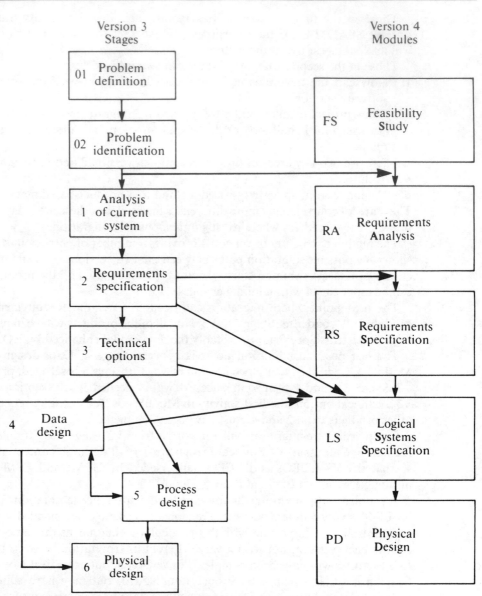

Figure 20.3

team to make the products from the V3 part conform to V4 standards. Every product that is going to be a deliverable for the User, or used as a subsequent input to a version 4 activity must be subject to Configuration Management.

Figure 20.4 shows the stages of V3 and V4 respectively, and the products from one that are needed to interface with the other, and where in the project they are generated.

The other features of V4, the Project Procedures and parallel activities, should not be a barrier or deterrent from converting from the older version to the new: all

Stage products

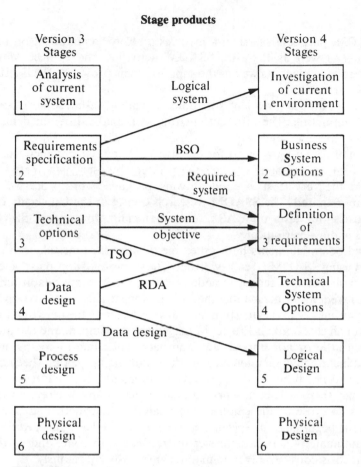

Figure 20.4

SSADM V4 has done is to make explicit what should be standard practices in any IS development environment. If the adoption of V4 leads to a Project Management structure being installed in the organization, rather than an *ad hoc* arrangement, that can only be to the good.

What might cause problems would be a simultaneous adoption of CASE technology. The problems would be cultural adjustment and an additional learning curve. If CASE is not already in use, this would make an excellent opportunity for introducing it. If CASE tools are installed, then the ramifications of introducing V4 need to be examined:

● Do the existing tools conform to V4 standards? Can they support the Product Descriptions?
● Can they be used in V3 and V4 projects until the absorption of V4 is complete?
● Can the quality criteria for products be specified and met using the tools?
● Can the tool carry out the cross-validation between the different SSADM techniques?

Summary

Use of SSADM, particular version 4, involves a heavy commitment on the part of management. As well as the core SSADM activities, the Project Management structure must be in place, as well as the support teams providing the interfaces with Project Procedures.

SSADM is used in a variety of situations, including in-house development and third-party contracting. The different environments make different demands on a practitioner's experience and skills.

Some form of computer support for the project is highly recommended, although the precise form of the support, and the products chosen, is a matter for local decisions. At the very least, a good drawing package, word-processor and data dictionary are required. The SSADM Research Centre at the University of Central England carries out tests on CASE tools that claim to support SSADM, and publishes its findings periodically.

Actual and potential clients have asked me if the extra time taken to produce a specification using SSADM is really worth it. The traditional approach of identifying a customer wish and proceeding to code is still active in some establishments, and such people take a lot of persuading that it is no longer a case of 'have coding sheet will travel'. My answer to the question must always be along the lines that I have tried to stress through this book: if Project Management is skimped, and the practitioner team is given little or no support, any approach will come to coding stage more quickly, and incur heavy maintenance overheads; if the project has been conducted responsibly, and has been fully supported, the timescales for delivery will be more protracted, but there will be a working system that is much closer to the User's requirements. Because of the emphasis on analysing and modelling User requirements from the beginning, the specification passed to the code generators—whether human or automated—should be easier to translate into code, and so reduce the development costs considerably, and maintenance costs enormously.

Appendix

The SSADM User Group is a body some 200 strong, with members drawn from government, industry, and academia. It has a number of special interest subgroups, such as a group for future development of the method, for current applications, for applications using distributed systems, for training, and many others. Any organization adopting SSADM for the first time would be well advised to join the User Group for support, to draw on other organizations' experience, and to be kept abreast of developments inside the SSADM community. The User Group holds a spring conference, and a residential autumn conference, and publishes a quarterly newsletter.

The address of the administrator of the User Group is given below. It is correct at time of writing, although obviously circumstances may change over the years. Two other addresses of interest are given, and one telephone number.

SSADM User Group
Mrs S. McGowan,
11 Burlings Lane
KNOCKHOLT
Kent   TN14 7PB

General SSADM
ISE Divisional Support Unit
Enquiries CCTA
ISE Division
Gildengate House
Upper Green Lane
NORWICH   NR3 1DW

SSADM Research Centre
University of Central England
Department of Computing
Perry Barr
BIRMINGHAM   B42 2SU

CCTA Publications
HMSO Publications Centre
Tel: 071 863 9090

Glossary

The glossary defines the terms and products that are used in SSADM. I do not classify the terms unless there is an ambiguity, so the range of expressions that follow refer to the Structural Model, SSADM techniques, products, measures, or interface elements.

Acronyms are entered against the full name, and are not given an entry in their own right. Thus 'Business System Options' has '(BSO)' printed against it. Readers who have only the acronym should search through the initial letter entries to find it.

acceptance testing criteria   The conditions that must be satisfied for the Users to accept the system as meeting their requirements. They define the final testing process before the system is handed over to the Users.

access path   This defines the entry point into the Logical Data Structure, and the navigation from entity to entity that is needed to perform a given piece of processing.

activity   Any action that transforms a product. There should always be a description held of each activity, defining the inputs, outputs, and participants and explaining what the transformation is.

Activity Network   A diagram that shows the sequence and dependencies of all activities. It is used as an aid to planning timescales and scheduling work and resources.

Analysis of Requirements   The outputs from Module 1, the Requirements Analysis Module. It comprises Current Services Description, Requirements Catalogue, User Catalogue, and Selected Business System Option.

application   That part of the business that is the object of the development work. It will have been defined initially in the Strategic Study.

Application Development Standards   Defines the standards by which the development of the current application is to be carried out. The standards are agreed by Project Management at the beginning of Stage 6; they are based on the Application Style Guide produced from Technical System Options.

Application Style Guide   A set of standards that covers aspects of the user interface; the standards apply to a particular development, although they are usually installation-wide. They are based on an Installation Style Guide, which should exist independently of this project, and are tailored to meet any particular requirements of this project.

attribute   A property of an entity, i.e., it describes the entity in some way. For each occurrence of that entity, the attributes are set to an appropriate value, the value being drawn from a domain. An attribute is also known as a data item. An example would be:

Entity	Attitude	Value	Domain
Course	Duration	10	Number of days, between 1 and 25
	Maximum Delegates	15	Number, between 5 and 20

An attribute may be 'optional', i.e., an entity occurrence may exist without a value for an attribute being given as soon as the occurrence is created.

batch   A grouping of events or functions that are performed within the same time-frame. An example would be the production of invoices. This would not be done piecemeal, one at a time, but in a batch together.

bottom-level process   A process on a Data Flow Diagram that cannot be decomposed to a lower level. It is marked on the diagram by an asterisk in the bottom right corner of the box.

Business System Option (BSO)   The mechanism for agreeing with the Users the scope of the functionality of the new system. BSOs are derived and selected in Stage 2 of SSADM, when the analysts prepare a number of possible scenarios for the new system. Each scenario addresses a different set of requirements from the Requirements Catalogue, and the Users, via the project board or similar body, will select one for development. The one chosen will be known as the Selected Business System Option, and will become the major input to Requirements Specification.

The analysts help the board make the selection by preparing details of comparative costs, benefits, impacts, timescales, and so on.

Capacity Planning   Not a core SSADM technique, but one of the Project Procedures that supports SSADM. CCTA have produced a publication on Capacity Planning explaining its use, and how it is applied.

Its purpose is to describe the hardware and software configurations needed to meet the new system's objectives and constraints. It looks at such areas as storage, access speeds, and performance prediction. It is also used as a tool to help develop the service-level requirements.

cardinality   *See* relationship degree.

Central Computing and Telecommunications Agency (CCTA)   The Government agency responsible for coordinating and overseeing the procurement and development of computing systems throughout government departments. Another responsibility is to recommend and implement standards in government computing. SSADM is one of those standards, and the CCTA has been the body responsible for commissioning its production, development, enhancement, and support.

To support the current version of SSADM, the CCTA is producing a set of publications which describe activities central to an IS development, but peripheral to SSADM itself. These guides define the principal management and technical interfaces to SSADM.

Command Structure   The control in a dialogue that specifies the navigational route that can be taken at the end of that dialogue. It can be terminated, or the User can move to a new one. The Command Structure permits navigation with menus or simply with commands.

common processing   Some elements of processing may be common to more than one function. A calculation, for example, may be used in more than one process; validation of inputs may be carried out in several similar inputs. Where such common processing is found, it must be documented in the Elementary Process Descriptions, with appropriate cross-references. They may also be carried forward in the Function Definitions.

component   A discrete part of a function, i.e., a part that can be separately identified. It is usually identified as one of the following: I/O process, database process, or common process.

composite data flow   A data flow on a Data Flow Diagram which can be decomposed at a lower level into two or more separate data flows.

configuration   A logically related set of products, requiring management and audit.

configuration item   A product that forms part of a configuration, ranging from a complete system specification to an algorithm, diagram, or piece of code. It is subject to Configuration Management, and as such is auditable.

Configuration Management   A set of techniques to manage a configuration. The techniques ensure that the configuration items are all produced according to specified procedures for a project, or local installation, and to the required quality criteria.

Context Diagram   A diagram that may be produced at the initiation of a project, or at Feasibility. It defines the boundary of the system and the major inputs and outputs that cross the system boundary. A Context Diagram may be the first product in the development of Data Flow Diagrams.

control flow   A management control that is exerted over any SSADM activity, whether Module, stage or step. The control may be to start the activity, stop it, or repeat it.

core SSADM   The five Modules of SSADM which make up the systems analysis and design activities.

Cost/Benefit Analysis   A method for providing an objective comparison between options. The costs of developing and operating each proposed system are calculated, and set off against the benefits of installing that system. It is the key supporting document for the options, both

BSO and TSO, and will usually carry most weight with the project board who will make the selection.

criteria for specification   A statement of the rules for specifying individual function components. These components may be procedural or non-procedural.

Current Environment Description   A full description, in SSADM terms, of the workings of the current system, both computer and non-computer aspects. Included is a list of identified shortcomings in its operations, and a list of requirements for the new system. If there is no existing system, the Current Environment Description will consist just of the requirements for the new system.

current services   All processing that now exists in the application area under investigation. The processing may be manual or computerized.

Current Services Description   The complete output from Stage 1 of SSADM. It comprises the details of the logicalization of the current system Data Flow Model, the Logical Data Model, and the identified problems and requirements.

Data Catalogue   A repository of definitions and descriptions for all attributes in the system, and all domains that give them their possible values.

data classification scheme   A means of documenting the data management facilities of the target implementation environment.

data flow   One of the elements of Data Flow Diagrams. The data flow shows a passage of data between two other elements, process, data store, or external entity. At the lowest level, the data flows are simple flows, but at the higher levels they may be grouped together as composite flows, for ease of drawing.

Data Flow Diagram (DFD)   A diagrammatic view of the functional view of the system under investigation. The notation comprises four elements: process, data store, external entity, and data flow. Its primary purpose is to communicate with the User the analyst's understanding of the scope of the system, the processing undertaken, and the interactions with the environment.

data item   An element of a data store that in some way describes, qualifies, or classifies that data store. A data item equates to an attribute on an entity. Each data item on a data store must have its equivalent on a corresponding entity.

data store   One of the component elements of a Data Flow Diagram. It represents any place in the system where data is stored or collected. It may be an electronic medium such as a disk/tape file, or it may be a manual storage method such as a filing cabinet, or card-index file. A batch of forms in a bulldog clip waiting to be actioned also counts as a data store. On a Data Flow Diagram, it is necessary to show each data store being accessed by at least one process, and usually more, both for creating the data, and for using it.

database management system (DBMS)   The mechanism for storing and retrieving data in a computer system. A DBMS allows data to be stored in such a way that every permitted application may use it, rather than have each application maintaining its own data files, with resultant duplication and loss of integrity.

DBMS Data Storage Classification   A means of analysing the mechanisms for storing and retrieving data on a database management system (DBMS). This is used in Stage 6, Physical Data Design.

DBMS Performance Classification   A record of the factors which impact on the performance of a DBMS. This is used in Stage 6, Physical Data Design.

detail entity   One of a pair of entity types. Entities participate in relationships, in which a single occurrence of one entity type (the master) is associated with several occurrences of another (the detail). A detail entity is denoted on the Logical Data Structure by a 'crow's foot' attaching it to the relationship line.

determinant   A term in Relational Data Analysis that denotes a key data item or group of items. The value of the other data items in the relation depend upon the value of the determinant. The determinant has a unique value for each occurrence of its relation.

dialogue   The occasion of a User Role performing a function on-line. The dialogue defines the exchanges that the User will have with the system via the terminal screen.

Dialogue Control Table   Produced during Dialogue Design (Step 510), Dialogue Control Tables show the possible navigation paths in a dialogue between the Logical Groupings of Dialogue Elements, highlighting the sequence of aspects of the dialogue.

dialogue element   A part of an input/output data flow. It may consist of just one item or it may contain several data items. The element is shown on the Dialogue Structure as a box.

Document Flow Diagram   The initial pass at scoping a system for drawing DFDs. The diagram shows the flow of documents around a system as flows from and to sources and recipients of each. Processing of these documents is not shown on this diagram.

domain   A complete set of possible values from which any occurrence of an attribute may take its actual value.

effect   The change caused to a single entity as a result of an event. The change will be to create an occurrence of the entity, to change the value of one or more attributes, including state indicator, or to delete an occurrence of the entity.

Effect Correspondence Diagram (ECD)   A diagram to show all the entities affected by an event within the system, and how these effects impact upon each other. It is produced in conjunction with Entity Life Histories in Step 360. ECDs are used to provide the access paths for Update Processing, used in Logical System Design.

element   Any unit or component of a product or activity.

elementary process   The lowest level processes on a set of Data Flow Diagrams.

Elementary Process Description (EPD)   The supporting documentation to describe the processing on a DFD. Every elementary process on the model has an EPD made for it, which describes the business activities that take place inside that process, and describes any decisions that need to be taken inside the process, using a tool such as a decision table, decision tree or structured English.

Enquiry Access Path   The path followed through a Logical Data Structure in order to satisfy a business enquiry. The Enquiry Access Path is developed with the Logical Data Model in Step 360, in tandem with the Effect Correspondence Diagrams.

Enquiry Process Model   A structure diagram (Jackson diagram) that defines the processing of an enquiry. Operations are shown on the diagram. It is created by merging the Enquiry Access Path with the output elements of the relevant I/O structure.

enquiry trigger   The data items that are input to define and trigger an enquiry.

entity   Something about which the system needs to hold information. There should be the potential for more than one occurrence of an entity, and each occurrence of that entity should be uniquely identifiable. They are of especial interest in Logical Data Modelling and Entity/Event Modelling.

Entity Description   One of the supporting documents for Logical Data Structures. Every entity in the system will be separately described, with all of the attributes and keys listed, as well as its purpose and use in the system.

Entity Life History (ELH)   A diagrammatic technique for showing the effect of events on an entity. The diagram portrays all the possible events that will affect any occurrence of the entity, for creation to deletion, with all the possible updates. The sequence in which these may happen models the business rules governing the entity. The notation is that of the structure diagram.

entity role   Denotes the situation where an event may affect more than one occurrence of an entity, but in each case the effect is different. Where this happens, the entity is said to have different 'roles', each role governing one of the effects. The roles must be represented on the ELH, as different processing is required for each role, e.g., an event that can delete one person's record may be responsible for creating a fresh occurrence, simultaneously. The roles in this case would be ⟨old⟩ and ⟨new⟩, and either deletion or creation would occur.

entity subtype   Two or more entity types that share significant characteristics, particularly key fields. An example might be Vehicles held by a company; the key field would be vehicle ID, but with different information required depending upon whether it was a Car or Truck. Common relationships, such as Service, would be made with the entity supertype (Vehicle); those with only the Truck would be connected with that subtype only.

entity supertype   *See* entity subtype. The entity type that embraces two or more subtypes. In the example above, Vehicle would be the supertype of Car and Truck.

Entity/Event Modelling   The joint production of ELH and Effect Correspondence Diagrams, in Step 360.

Estimating   One of the Project Procedures that support SSADM. Producing estimates for activities and costs is a continuing activity through the project, requiring constant revision. Initial estimates are based on identified activities; these activities are weighted according to a number of factors, such as the complexity of the system, the experience of the practitioners, resource availability, and so on. Estimates are made for each activity, whether Module, stage, or step.

event Something which happens in the world that causes a change in the value or status of a data item or entity in the system. It can be shown as a data flow responsible for updating a data store on a DFD.

Event/Entity Matrix A technique to examine all the events that affect data in the system, with reference to all the entities affected by each event. It provides cross-references between LDS and ELH by ensuring that all entities are created, modified, and deleted by identified events; conversely, it ensures that every identified event affects at least one entity.

external entity A body outside the system (possibly a statutory organization such as the Inland Revenue, another system in the organization, an individual, or a group of people). The external entity communicates with our system by means of data flows in or out of the system. The entity may be either source or recipient of the flows, or both. It is shown on the DFD as an oval with the name of the external entity inside, and a unique identifier.

Feasibility Study The activity carried out in the first Module of an SSADM project. Feasibility is intended to define the scope of the project, and the direction it will take. During Feasibility two basic questions are addressed: can the objectives of the study be met, i.e., is it *technically* feasible? And is there a sound business case for meeting the objectives?

The study is carried out using a variety of techniques, including SSADM techniques like Data Flow Modelling and Logical Data Modelling. The report produced at the end of the Feasibility Study represents one of the User options points in SSADM, where the direction for the next part of the project is determined.

first-cut Data Design (first-cut design) The Logical Data Model is subject to a transformation based on certain rules-of-thumb. These are independent of any given implementation environment, but should subsequently be adaptable to any specific product. It should be close enough to the final design for timing and sizing exercises to be carried out.

There are two versions of the first-cut Data Design: The implementation-independent design, followed by a first-cut design according to the supplied rules of the chosen environment. The former is often omitted, depending on the experience of the designer, or the requirements and complexity of the selected DBMS.

fragment A defined processing element with a specific purpose and defined inputs/outputs. It may correspond to an operation, or to a function component. It may support a process, a data group, or a screen message. Fragments are identified and defined during the compiling of the FCIM.

function A User's view of a piece of system processing; the processing as viewed supports a business activity rather than a system or operations activity. The function is broken into smaller operational components for processing purposes, such as input components, output components, and database read/write components. Each function is classified according to whether it is an update or enquiry function, whether it is on-line or off-line, and whether it is triggered by User input, or by the system itself.

Function Component Implementation Map (FCIM) A decomposition of all functions into the component elements to be implemented. The components are identified in Function Definition, and cover the following: superfunctions, functions, I/O processes, database processes, and common processing.

Function Definition A technique to identify and define the functions carried forward for Physical Design. It takes place in Step 330.

functional requirement A requirement from the User for a particular service from the system. The service is a business request, and may be an output form, an update process, an ad hoc query facility, or a periodic report.

Impact Analysis An examination of the effects on an organization of either a Business System Option or Technical System Option. It is one of the supporting documents produced for each option presented, and covers issues of organization, staffing, and work practices.

information highway   The mechanism by which the products of each SSADM activity are submitted to Project Management for Quality Assurance review. All SSADM products are passed to the highway rather than to the next activity, and must be reviewed. Any Configuration Management procedures in use operate within the information highway.

Input/Output Description (I/O Description)   An SSADM document to record all data about data flows on the Data Flow Diagram. It is completed in Steps 130, 150, and 310.

Input/Output Structure (I/O Structure)   The collected and documented data items for a given function. The I/O Structure comprises I/O Structure Diagram and I/O Structure Description.

key data item (or key)   An attribute whose value uniquely identifies a specific occurrence of an entity. The values of all other attributes in that entity depend upon the key. In Relational Data Analysis the key is often referred to as a determinant (as its values 'determines' all other attribute values).

   There are several categories of key, as defined in the chapters on Relational Data Analysis and Physical Data Design.

Logical Data Flow Diagram   The description of the current physical environment, after all references to physical and contingent factors have been removed. The aim is to understand the business logic underlying the current practices, rather than simply to model the procedures in place, regardless of their validity. Typically, references to constraints caused by geographical considerations, machine-related procedures (e.g., photocopying, data prep.), office politics, or office procedures are removed from the Current Physical DFD, leaving just the business logic. This takes place in Step 150.

Logical Data Model (LDM)   The collected Logical Data Structure and supporting documentation (Entity Descriptions, Relationship Descriptions) that define the data structure of the target system. The Overview LDM contains the Overview LDS only, without the supporting documentation. That is added as the LDS is expanded.

   It is derived and built up in Steps 010, 020, 110, 140, 320, 340, 360, and 520.

Logical Data Store   The repository on the Logical DFD for all data items belonging to one subject. Unlike the physical data stores, where one topic, e.g., orders, could be held in many formats and many places during the life of any one occurrence.

Logical Data Store/Entity Cross-reference   A cross-validation between the Logical Data Model and the Logical Data Flow Diagram. The cross-reference consists of two columns; Logical Data Stores are listed in one column, and against it in the other is drawn the entity or groups of entities that correspond to it, i.e., that have as attributes those data items held in the data store. All entities on the LDM correspond to one and only one Logical Data Store.

Logical Data Structure   One of the three 'views' of the system. It shows the information requirements of the organization by identifying entities and the relationships between them, in terms of the business functional activities.

Logical Design   The output from Stage 5, Logical Design. It embraces the Required System Logical Data Model, the logical processing, and the Requirements Catalogue. It serves as the input to Physical Design.

Logical Grouping of Dialogue Elements (LGDE)   A grouping together of dialogue elements, usually for operational reasons. It is identified in Step 510, Dialogue Design, by lassoing together the appropriate elements at the leaves of the Input/Output Structures.

Logical Process Model   The collection of all processing details incorporated in Logical Design.

logicalization   The process of converting the DFD of the current environment from the physical description to one reflecting the business logic only. The process involves rationalizing the processes and data stores.

master entity   The entity at the 'one' end of a relationship in an LDS. Of the two entities in a relationship, one is associated with one and only one occurrence of the other, while the other may be associated with more than one occurrence of the first. The entity that may occur many times is known as the detail entity and the one that occurs once only is the master entity.

Menu Structure   A diagram that shows a hierarchy of menus within an on-line system, with the possible navigation paths between menus and dialogues.

Module   The significant division of SSADM activities, for management purposes. An SSADM project is composed of five Modules, each of which is made up of one or more stages. Each Module produces a defined set of products, and performs a defined set of activities. The Module is deemed to be complete when the products are completed to predefined quality criteria.

non-functional requirement   A requirement of the new system that describes the service levels or security levels that it must meet. It meets a performance need, rather than a fundamental business need. It may be a requirement for particular response times, or it may be to do with the mean time between failures (MTBF) or recovery time from failure.

normalization   A synonym for Relational Data Analysis, a way of transforming unstructured data into minimal logical groupings, such that each attribute is grouped only with its sole determinant and other attributes that share that determinant. The aim of normalization is to ensure that every entity is in the condition known as Third Normal Form.

off-line function   A function that is performed by the computer without terminal interaction with the User. The trigger for the function may be input on-line, or in batch mode, but all the database processing will be carried out subsequently without reference to the User.

on-line function   A function that is performed by the computer while the User is at the terminal; the system and the User communicate with input and output messages while the function is being executed, the contents of the input messages being influenced by the preceding output messages.

operation   Discrete pieces of processing which together constitute effects. They are identified during Entity/Event Modelling, and are entered on the Entity Life Histories. Subsequently they are expanded on the Update Process Models.

optimization   Tuning of a database model to enable it to meet the performance requirements of the specification. Optimization is a broad term to cover several kinds of activities, that range from changing the Logical Data Model, to altering the physical placement of data items on the disks, to amending the way that the program logic works.

Outline Current Environment Description   A product of Feasibility Study, it describes the current services in the environment, and the problems associated with their provision. The description is at a high level, rather than at a detailed level, and includes overview DFDs and LDS of the current system.

Outline Required Environment Description   Another product of the Feasibility Study, this describes in broad terms the scope and requirements of the new system. As well as text description, it may include overview Data Flow Diagrams and Logical Data Structure.

parallel structure   An element of Entity Life History notation. It denotes possible updates to the entity occurrence that can happen at any time during its existence, without a predetermined sequence, and without affecting the business cycle impacts on the entity. Reference data such as 'Address' on a customer record will be affected by events on parallel lives.

performance requirements   A set of non-functional requirements that define the performance of the new system. The areas covered include response times for on-line systems and turn-round time for off-line systems, as well as recovery specifications.

Physical Application Specification   The output from Physical Design Module. It is a collation of all the documentation that makes up the technical design.

Physical Data Design   The database definition which is to be implemented as a result of the optimization of the Physical Data Model.

Physical Data Model   The Logical Data Model after it has been transformed in Stage 6 to a Physical Model, by applying first a non-specific set of rules for transforming it, and then applying the rules of the target DBMS to that model. Until it has been optimized in one of several possible ways, it is unlikely to meet the performance requirements, and so is not yet a Physical Data Design.

physical environment   The selected environment on which the new system will be implemented. The term embraces DBMS, with its classifications of storage and performance. Also included are details of hardware products, software development environment (i.e., 3GL, 4GL, and so on) and other software such as TPMS, operating system, etc.

Physical Function Specification   A full description of the processing requirements of the new system.

physical key   A device to locate the target record on the physical disk. It is also known as a 'pointer'.

problem definition statement   The definition of the User requirements for the new system. It is produced during the Feasibility Study, and incorporates charts and diagrams.

procedural model   A structure diagram with operations and conditions appended. It is used as the formal definition of the procedural processes.

process   An element of a Data Flow Diagram. It records any transformation or manipulation carried out on a data flow or on data items.

Process Data Interface (PDI)   Produced during Physical Design, it shows how the logical processes that access the LDM are mapped onto the physical processes that access the physical database. If there is a one-to-one correspondence between logical and physical accesses, then there is no need for a PDI; if the two do not match exactly, the PDI shows how the mismatches can be resolved in terms of physical access. If a non-procedural language such as SQL is used at the site, the PDI can be implemented in just one or two lines. The aim of the PDI is to allow the designer to implement the logical Update and Enquiry Processes as programs, independently of the physical database structure.

Process/Entity Matrix   A tool to help group bottom-level processes during logicalization. It is also used to perform a cross-validation between LDS and DFD, by ensuring that every entity is affected by at least one creation and one deletion event.

Processing Specification   The output from Step 360. It comprises the following assembled products: User Role/Function Matrix, Function Definitions, Required System LDM, Entity Life Histories, and Effect Correspondence Diagrams. The development of this product may identify errors in the component parts (e.g., there may be some events on the ELHs never reflected in the Effect Correspondence Diagrams). When this is the case, a revisiting of the relevant products and techniques is called for.

Processing System Classification   Classifies the details of the processing environment used for implementation. It frequently defines the development environment, too.

product   Any item of documentation, software, or hardware that is produced during an SSADM project. The product itself may be made up of component products, such as the Process Specification. The Product Breakdown Structure classifies all products under three broad headings: management products, technical products, and quality products.

Product Breakdown Structure   A hierarchic structure that itemizes all SSADM products, and categorizes them under three broad headings.

Management products encompass such documents as Project Initiation Document, Application Development Standards, Highlight Reports, Project Plans, Module and Stage Plans, and Post-implementation Review.

Technical products include application products, operations products, education products, and release packages. The products from systems analysis and design activities come under the heading of application products.

Quality products refer to all aspects of the continuing Quality Assurance process. The products include invitations to quality control reviews, records of each review, and records of follow-up action required.

Product Description   A specification for every product listed on the Product Breakdown Structure. Project Management are responsible for completing these as part of the input to every activity. Information listed in the descriptions include derivation, composition, quality criteria, and external dependencies.

program   A set of codes that fulfils the requirements of one or more functions.

project   A set of activities that together present the following features: a defined and unique set of technical products that meet the business needs; a corresponding set of activities to produce those products; nominated resources; plans; objectives; and deadlines.

Project Management   A set of techniques and an organizational structure to plan, monitor, and control the conduct of a project. It is concerned with the monitoring of the project resources and the observation of any constraints.

Project Procedures   The tools of Project Management that support the activities of core SSADM. They include such activities as Risk Analysis, Capacity Planning, Estimating, and

Configuration Management. They are handled by specialist teams outside the SSADM practitioner team, and interface via Project Management.

prototyping   A method of providing the User with a simulation of how the system will work. The simulation is modelled and demonstrated on a specialist tool that imitates the features of the implementation environment. It is used in SSADM to confirm or clarify the User requirements.

Prototype Pathway   A document showing the sequence and combinations of dialogue elements and screens demonstrated in a prototyping session with the User.

quality   A measure applied to all SSADM products to show that they are fit for the required purpose. Every product is classified according to standard criteria, one set of which is the quality criteria. If the product meets these criteria, it may then be accepted as part of the project documentation.

quality assurance   A set of mechanisms that collectively guarantee that all products from a project meet the required standards.

Quality Control   The process of ensuring that the required quality criteria are built into the products. This can be done by various forms of inspection or review during the course of production and before acceptance of a product.

quality criteria   Prescribed characteristics of a given product; these characteristics are specified before the production begins, and are passed to the practitioner team via the information highway.

quit and resume   A feature of Entity Life Histories. It is a device to show that particular events can cause a change in the prescribed sequence, in certain circumstances. It is a notation that describes how processing is stopped at one point and resumed elsewhere.

random event   An event which may occur at any time during the life of an entity, rather than at a predetermined stage in its life.

relation   A group of data items under its logical determinant. It is represented as a table of entries, each row entry being an occurrence of that relation. In Relational Data Analysis each relation is regarded as an entity for the purposes of validating the Logical Data Model.

Relational Data Analysis (RDA)   A technique for analysing data to its logical groupings. A set of 'raw', i.e., unstructured, data is reduced through a set of transformations (normalization) to basic data groups in which each item is grouped with its sole and full determinant. The effect of performing RDA, as a step in data design, is to reduce data redundancy and gain flexibility. During the normalization process, it is important that information is never lost, i.e., that all the original connections between data items are maintained. The output from RDA is a set of relations that are in the state known as Third Normal Form.

relationship   An association between two entities. A relationship that is identified on a Logical Data Model must hold true for all occurrences of those entities.

relationship degree (cardinality)   The class of number of each entity taking part in a relationship. At one end of the relationship there may be one and only one occurrence of an entity type. This entity type is the master in the relationship. At the other end there may be many occurrences of that entity type; that is called the detail. Relationship degree classifies the ends of the relationship as being one-and-only-one, and many. The degrees, or cardinality, are described in the following terms:

- $1:1$ one to one
- $1:m$ one to many
- $m:n$ many to many

Required System Data Flow Model   The version of the Data Flow Model that describes the processing in the new system.

Required System Logical Data Model   The version of the Logical Data Model that shows the structure of the data for the new system, and the information requirements.

requirement   A feature of the new system that has been requested by the User. The requirement may be functional (i.e., to perform the application environment's primary activities) or non-functional (i.e., service requirements such as specified response-times for on-line activities, or recovery/security requirements).

Requirements Catalogue   The central document that lists all requirements from the new system. It is started at the beginning of the project, and is maintained through all stages until Physical Design is finished.

Requirements Definition   A procedure running through the early parts of an SSADM project. There is discussion over whether this counts as a technique or a procedure. It focuses on the provisions of the required system which will meet the User's business needs. In the later parts of the project the identified requirements are resolved by techniques such as Function Definition.

Resource Flow Diagram   A component part of a Data Flow Diagram. Its purpose is to show the physical movement of goods in a system, rather than the movement of information. Its function is to help with the initial scoping of the system, and to help identify flows of data in the early stages of the analysis.

response time   The time taken for the system to respond to a User input in an on-line activity.

Risk Analysis   A set of procedures to identify risks to aspects of the system. These aspects may be hardware, security, software, physical installation, or any area where a threat is perceived. As well as identifying the risks, the practitioners must assess them, and prepare contingency measures to counter the risk. The aim is to reduce the risk to an acceptable level, but at an acceptable cost.

service-level requirement   A requirement for the new system, but at an operational level, rather than a business function level. It describes the expected, acceptable level of service in terms of response time and recovery times.

Space Estimation   A special form to assess the storage requirements of the data design for the target implementation environment.

Specification Prototyping   A technique employed in SSADM to validate selected User requirements by prototyping certain dialogues and procedures. It is not used as an incremental way of building a system, only as a validation exercise with the Users.

stage   A unit of activity within an SSADM Project, with defined input and products. It is a subunit within a Module.

state indicator (SI)   A feature in entity Life History analysis. An SI consist of a non-functional attribute (single-digit numeric) in an entity. Every event that affects that entity causes a change in the value of the SI. A birth event sets it initially to value = 1, and a death event sets it back to value = null.

The SI can be used as a validation or referential check: if an SI is not set to a valid value for the event, the process will be aborted and an integrity error flagged. In this way, all the updates to the entity can happen only in the correct sequence.

step   A subunit of activity within a stage. Like a stage it has its prescribed inputs and products.

Structural Model   The model of SSADM that describes the Activity Descriptions and architectural description to define the activities carried out in the SSADM part of a project.

structure clash   A situation during Logical Process Modelling, when the two structures being merged prove to be incompatible. There are three recognized forms of structure clash: ordering clash, where the data items in the input structure occur in a different order from the output structure; boundary clash, where not just items, but groupings of items, are different on the two structures; and interleaving clash, where, in the input structure, the data items from different entities are mixed together, rather than all data items from entity 1, then entity 2 and so on.

Structure clashes will probably not be resolved until Physical Design. The resolution may be provided by the implementation platform, or may require an extra process, such as a sort process, to bring input and output diagrams into compatibility. The CCTA publication on the 3GL interface describes in detail different approaches to resolving structure clashes.

structure diagram   A diagrammatic notation for structures or procedures, shown in terms of sequence, selection, and iteration. These three basic constructs are sufficient to model the structures richly. They are based on Jackson diagrams, used in Jackson structured programming (JSP), and Jackson structured design (JSD).

success unit   A set of processing which must succeed or fail as a whole for each input. If there is a failure before the success unit is completed, and the processing aborts, the system is restored to the state it was in before the success unit began. The success unit may be a complete update, a complete enquiry, an enquiry–update pair, or a predefined number of database accesses in an enquiry, depending upon the function's requirements.

system   In SSADM terms, the totality of the technical output from the project. The system exists outside the boundary of the project, and exists—as a result of the project—long after the project is complete. The system will have existed before the project as well, if it is an enhancement project.

Technical Environment Description (TED)   The description of the technical environment after the selection of the Technical System Option. The description is a major input to Physical Design. The TED contains a configuration diagram of the hardware, and gives details of the type of device, quantities, and locations.

The documentation in the TED describes the hardware, in terms of standards, networking requirements, back-up, upgrade arrangements, installation serviceability, and maintainability. Software is described in terms of the type of DBMS, utilities required, application packages, method of developing the applications, the development environment, and the number of applications programs required.

The TED also provides sizing details on data for storage forecasts, and processing times for all the identified functions.

Technical System Option (TSO)   TSOs describe the technical implementation of the Requirements Specification. It is prepared and selected in Stage 4, Technical System Options. The stage has two steps: 410, Define Technical System Options, and 420, Select Technical System Option.

The option/selection procedure is similar to that for the BSO. In addition, the team preparing the TSO need to receive information from the information highway on Risk Assessment and Capacity Planning. The documentation supporting the TSO is considerably more detailed than that for the BSO and, following the adoption of one solution, tenders can be invited on the strength of those documents and figures.

Third Normal Form (3NF)   The output from Relational Data Analysis. A relation that is in Third Normal Form is expected to be in the most logical grouping, not to be subject to update anomalies, and to give the least data redundancy. The way to achieve Third Normal Form is to examine all Second Normal Form relations for dependencies between non-key data items.

Timing Estimation form   A form designed during Stage 6, Physical Design. It is designed after the target DBMS or file system has been classified, and is intended to show how long a defined transaction will take to be performed, according to the database design and the facilities of the target DBMS.

transient data store   A store of data that is held temporarily before being processed, and is then deleted. Such data is not likely to be included on a Logical Data Structure. These stores are mostly used for administrative convenience, and frequently vanish from the Data Flow Diagram at logicalization.

Universal Function Model   A graphical representation of an SSADM function, showing how the SSADM products from earlier stages are built into the components of the function for processing.

Update Process Model   SSADM structure diagram that defines the processing sequence of an update transaction. It is a synthesis of Entity Life Histories, with the associated Effect Correspondence Diagrams, and lists the operations on the database that are carried out during the processing.

User   The person/job that will make use of the IT system to perform a business task.

User Catalogue   A document identifying the Users of the on-line components of the system. It incorporates job titles and job descriptions.

User Role   A collection of Users who share particular tasks or functions.

User Role/Function Matrix   Used in the identification of dialogues, the User Role/Function Matrix is a grid that marries up User Roles with on-line functions. Where there is a match between the two, a dialogue is required.

Solutions to exercises

Chapter 8

1. (a) See Fig. S.1. The entity Loan acts as a link entity between Member and Copy. While link entities are not required in the current description to resolve many-to-many relationships, in this case it is required as a real entity in its own right, even though the environment description does not mention Loan at all.

   Blacklist is mentioned, but does not qualify as a candidate entity, because there is only one occurrence. There are many entries, so Blacklist Entry might be regarded as an entity. Further investigation reveals that it is an attribute of member, a blacklist indicator, so it is not shown on the diagram. It will be shown on the model, though, in the Entity Description of Member, and with its own Attribute Description.

   Fine is in a one-to-one relationship with Loan. This is acceptable for a current system model, but in a Required System LDM it would have to be resolved. The resolution could be either to include it as an attribute of Loan, or to treat that relationship as a one-to-many. Relational Analysis in Step 340 may help the decision.

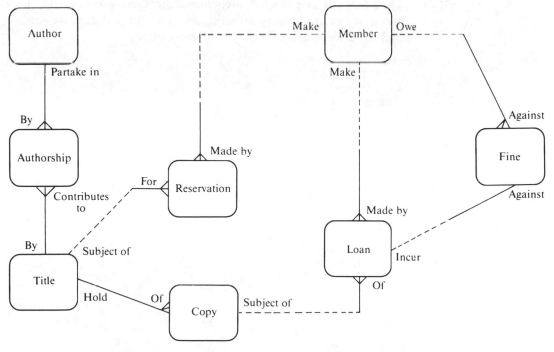

Figure S.1

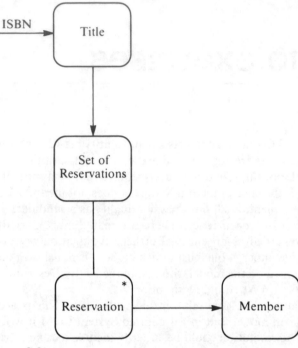

Figure S.2

(b) (i) Figure S.2. is a simple EAP, involving a short navigation across three entities. As we are looking for all reservations for that title, we need to build the extra structure box Set of Reservations.

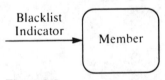

Figure S.3

(ii) Figure S.3 is an even simpler EAP, involving one entity and one search criterion. An EAP is no less good for being simple!

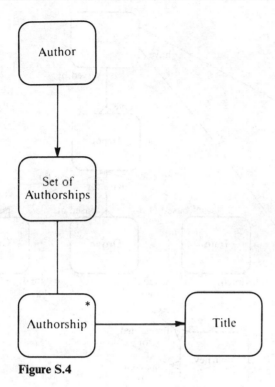

Figure S.4

(iii) Figure S.4 is similar to (i), involving a simple master to Set of details, then for each detail, its master.

2.   (a)

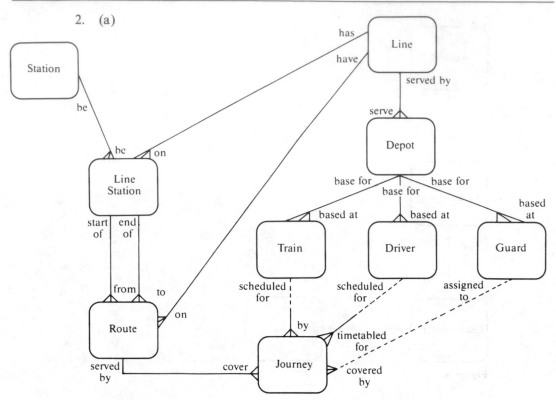

Figure S.5

Figure S.5. is a more complex diagram. The main point to notice is that the relationship between Line Station and Route is a double one: the start of a Route, and the end of a Route. In such a case, each relationship must be drawn on the model, and each relationship will have its own pair of Relationship Description forms.

   There is one relationship which has an optional master: Journey to Guard. Most of the other relationships are mandatory.

(b) The EAP (Fig. S.6) is again straightforward, from master to detail, preserving the hierarchic shape of that portion of the LDS.

(b)

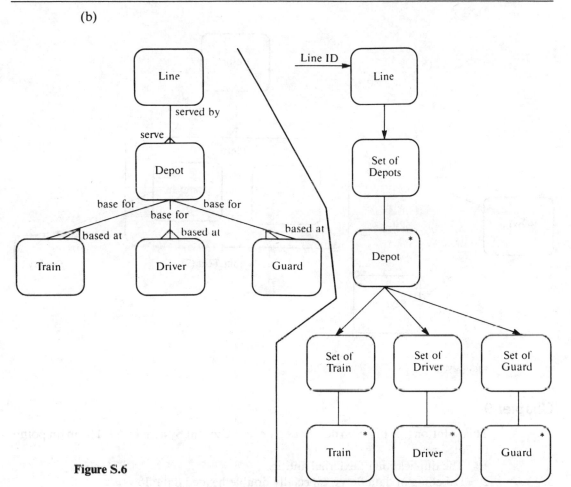

Figure S.6

3. Figure S.7 demonstrates the use of entity sub typing. The supertype is Plot, for which Bookings are made. Each Plot, though, may be either Tent, or caravan. If it is a Caravan, then there is an extra relationship, with Facility. I have given the solution in two different notations. Which one is used in any project is a matter for the Installation Style Guide and local standards.

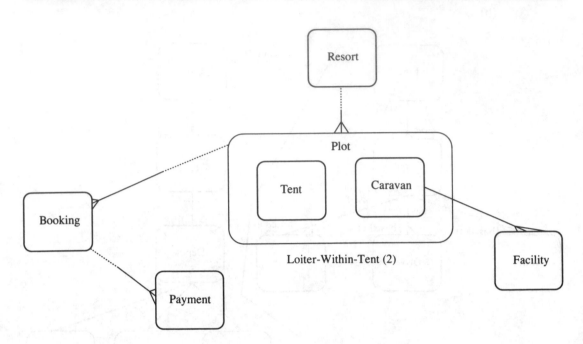

Figure S.7

Chapter 9

This solution (see Fig. S.8) describes a Level 1 Current System DFD. The main points to note are:

1. The duplication of external entities.
2. Labelling of data flows, especially double-headed data flows.
3. The bottom-level process (Process 2).
4. The data flow into D3 Provisional Bookings from Process 2: although no data is being written to the data store, it is being amended by a deletion. Therefore, the flow must still show the store being updated.

Process 2 has an Elementary Process Description. The other processes are decomposed to identify the contributory tasks and data stores accessed. For instance, Process 3, Pay Agent's Commission must look up agents' contracts, tax tables, etc., in calculating the amount to pay, as well as simply totting up the business they have put in Old Krate's direction.

Because Process 3 is carried out monthly, the trigger is shown as a data flow from the data store, rather than a flow across the boundary.

A Data Catalogue will be opened for all the data items in the data stores.

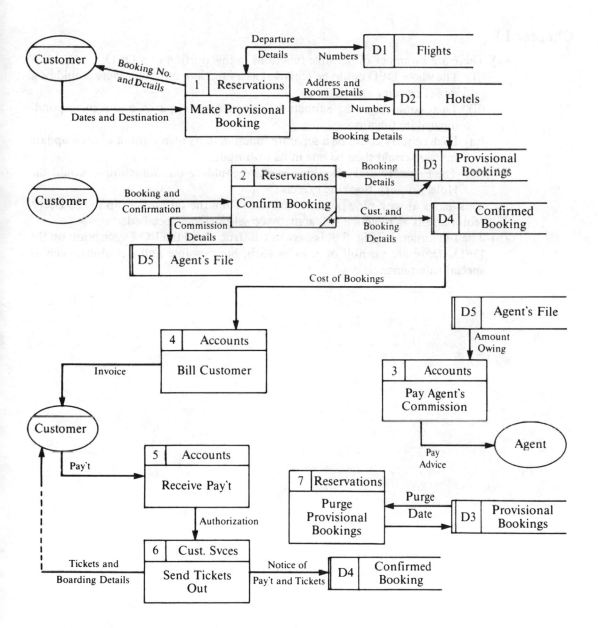

Figure S.8

Chapter 10

(a) There are a number of possible functions in this portion of a DFD:

    (i)   The whole DFD could be regarded as a function, although there would be a mix of enquiry and update.

    (ii) There could be a single function for both enquiry paths, and a corresponding update function.

    (iii) Each enquiry could be a separate function in its own right, and each update function could then be one in its own right.

    (iv) The Flight Enquiry/Booking exercise could be one function, as could the Hotel Enquiry/Booking exercise.

This leaves us with nine possible functions from the DFD. It is up to the User's requirements as to which will actually be selected and specified.

(b) The I/O Structure (Fig. S.9) is easy to construct from the I/O Description on the DFD. There are no null options as such, but possible null iterations, such as special requirements.

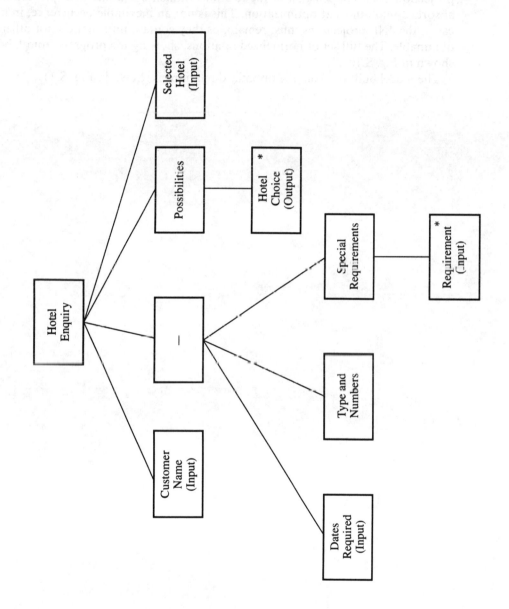

Figure S.9

Chapter 11

This exercise extends to Fourth Normal Form. There are no instances of a 5NF projection. As it happens, all of the relations broken out in the 4NF projection are absorbed into others at optimization. This is not an inevitable occurrence; in many cases, the 4NF projections must remain, as they contain information not otherwise obtainable. The full set of normalized relations, showing the progress from UNF, is shown in Fig. S.10.

The model built up from the optimized relations is shown in Fig. S.11.

System: Hospital rotas

UNF	1NF	2NF	3NF	BCNF	4NF	5NF	Optimized
Ward Name Date Ward Type No. Beds Sister Nurse No. Nurse Name Nurse Exp. Pat. No. Pat. Name Bed No. Date/Adm. Ded	Ward Name Date Ward Type No. Beds Sister Ward Name Date Pat. No. Pat. Name Bed No. Date Admin. Ded Ward name Date Nurse No. Nurse Name Nurse Exp.	Ward Name Date Sister Ward Name Ward Type No. Beds Ward Name Date Nurse No. Nurse No. Nurse Name Nurse Exp. Ward Name Date Pat. No. Bed No. Pat. No. Pat. Name Date Admin. Ded	Ward Name Date * Sister Ward Name Ward Type No. Beds Nurse No. Date * Ward Name Nurse No. Date Ward Name Nurse No. Nurse Name Nurse Exp. Pat. No. Date * Ward Name Bed No. Pat. No. Pat. Name Date Admin. Ded	Sister Date * Ward Name	Ward Name Date Nurse No. Date * 4NF relations projected from	No 5NF relations Nurse No. Date Ward Name	**Sister** Ward Name Date * Sister **Ward** Ward Name Ward Type No. Beds **Ward Staff** Nurse No. Ward Name Date Ward Name **Nurse** Nurse No. Nurse Name Nurse Exp. **Patient Ward** Pat. No. Ward Name Date * Ward Name Bed No. **Patient** Pat. No. Pat. Name Date/Adm. Ded Sex * Consultant No.

'Sister' turns out to have 'Nurse No. as its key.

Relationships must be noted if Nurse is in role of Sister.

Figure S.10a

365

System: Hospital rotas

UNF	1NF	2NF	3NF	BCNF	4NF	5NF	OPTIMIZED
<u>Consultant No.</u>	<u>Consultant No.</u>	<u>Consultant No.</u>	<u>Consultant No.</u>	<u>Pat. No.</u>	No 4NF relations	No 5NF relations	**Bed Allocation** (Ward Name)
<u>Date</u>	<u>Date</u>	<u>Date</u>	<u>Date</u>	<u>Consultant No.</u>			(<u>Bed No.</u>)
Specialism	Specialism						<u>Date</u>
Pat. No.		<u>Consultant No.</u>	<u>Consultant No.</u>	Ward Name			* Pat. No.
Pat. Name	<u>Consultant No.</u>	Specialism	Specialism	<u>Date</u>			
Ward Name	<u>Date</u>			<u>Pat. No.</u>			**Consultant Duty**
Bed No.	Pat. No.	<u>Consultant No.</u>	<u>Consultant No.</u>	Bed No.			<u>Consultant No.</u>
Sex	Pat. Name	<u>Date</u>	<u>Date</u>				<u>Date</u>
	Ward	<u>Pat. No.</u>					
	Bed No.		<u>Consultant No.</u>				**Consultation**
	Sex	<u>Date</u>	<u>Date</u>				<u>Consultant No.</u>
		<u>Pat. No.</u>	<u>Pat. No.</u>				<u>Date</u>
		Ward					<u>Pat. No.</u>
		Bed No.	<u>Date</u>				
			<u>Pat. No.</u>				**Consultant**
		<u>Pat. No.</u>	Ward				<u>Consultant No.</u>
		Pat. Name	Bed No.				Specialism
		Sex					
			<u>Pat. No.</u>				**Sister**
			Pat. Name				<u>Sister</u>
			Sex				<u>Date</u>
							* Ward Name

Figure S.10b

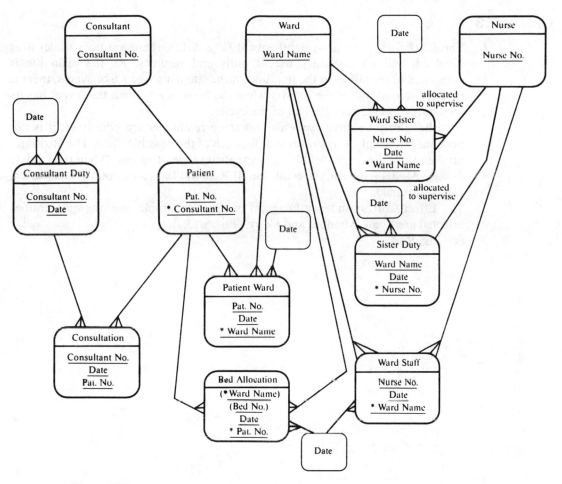

Figure S.11

Chapter 12

1. The ELH from loan is straightforward (Fig. S.12); there are no parallel lives involved, and no necessary use of quits and resumes. As the main life is composed of selections in the iteration, rather than a sequence of events, there is no need to quit out of the activity when the book is returned; that event has its own natural place near the end of the cycle.

   Although up to three renewals and three reminders are permitted, it is not necessary to specify a separate event for each of the possible three. The attributes on the record are single fields with a specified range of values. When the Update Process Model is drawn, the condition 'If Renewal Indicator not > 3' and so on, can be specified.

2. The Effect Correspondence Diagram is, again, a simple one: the only entities affected are Loan, Member and Copy. (Fig. S.13.)

3. See Fig. S.14.

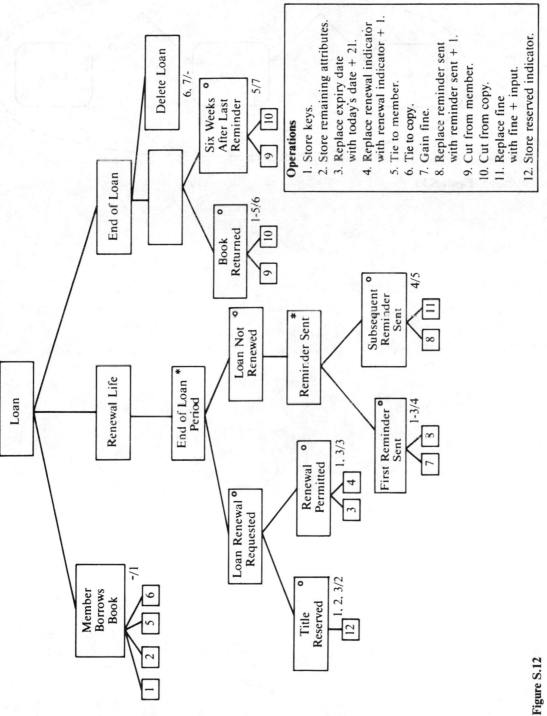

Operations
1. Store keys.
2. Store remaining attributes.
3. Replace expiry date with today's date + 21.
4. Replace renewal indicator with renewal indicator + 1.
5. Tie to member.
6. Tie to copy.
7. Gain fine.
8. Replace reminder sent with reminder sent + 1.
9. Cut from member.
10. Cut from copy.
11. Replace fine with fine + input.
12. Store reserved indicator.

Figure S.12

369

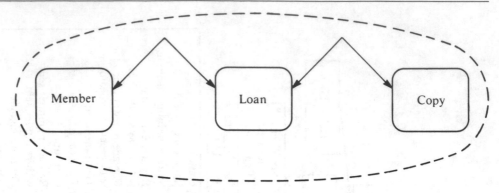

Figure S.13

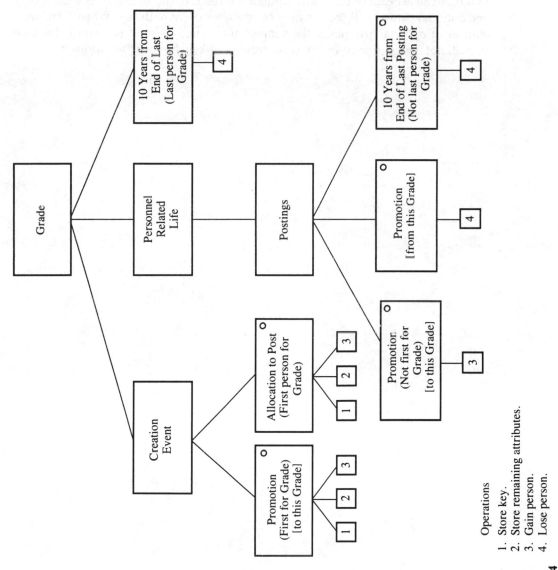

Operations

1. Store key.
2. Store remaining attributes.
3. Gain person.
4. Lose person.

Figure S.14

Chapter 13

The LGDEs for Cancel Booking are straightforward (Fig. S.15). Only the last one, defining Refund, is optional. Even though there are two iterations, Date and Seat-within-Date, neither is optional, at least one of each will be input, as they comprise the Booking which is being cancelled. Therefore, they must be mandatory LGDEs.

The Dialogue Control Table (Fig. S.16) shows the default pathway taking in every LGDE, as 80 per cent of cases will stipulate a refund. In this instance, as there is only one optional pathway, there can only be one alternative pathway. When there are a number of optional groupings, the number of alternatives must necessarily increase as well, not always one-to-one, but according to likely business situations.

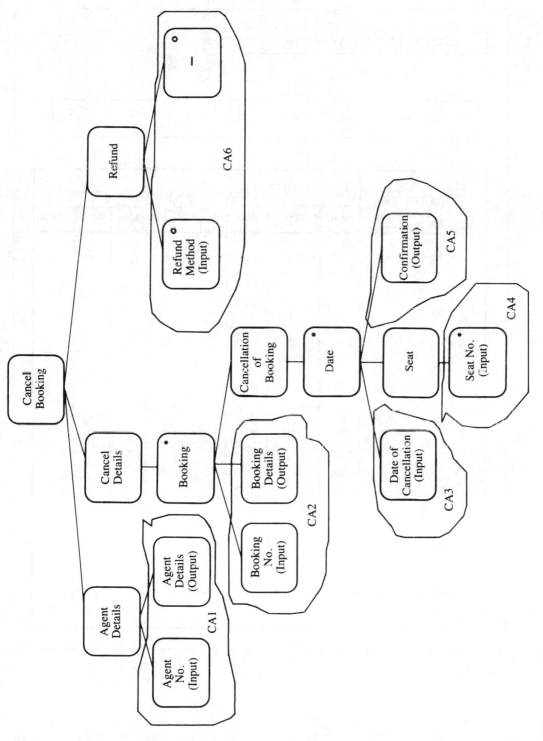

Figure S.15

Dialogue Control Table

Dialogue name: Cancel Booking

LGDE	Occurrences			Default pathway	Alternative pathways		
	Min.	Max.	Ave.		Alt 1	Alt 2	Alt 3
CA1	1	1	1	X	X		
CA2	1	1	1	X	X		
CA3	0	1	1	X	X		
CA4	0	20	9	X	X		
CA5	0	5	3	X	X		
CA6				X			
Percentage path usage				70	10	10	10

Figure S.16

Chapter 17

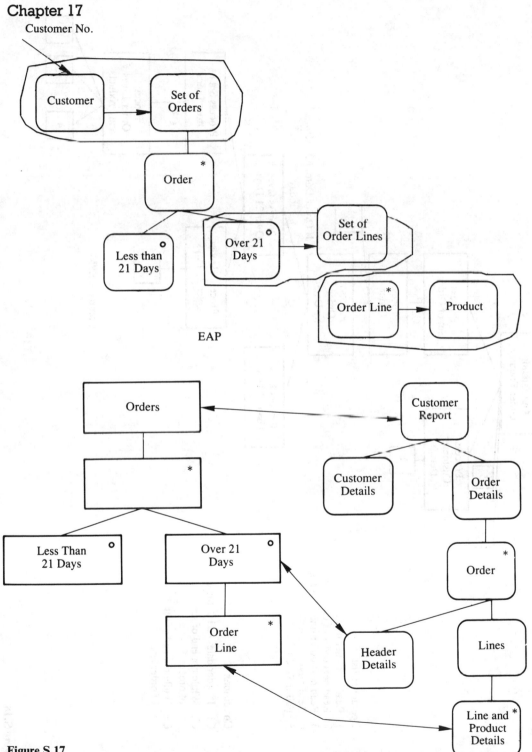

EAP

Figure S.17

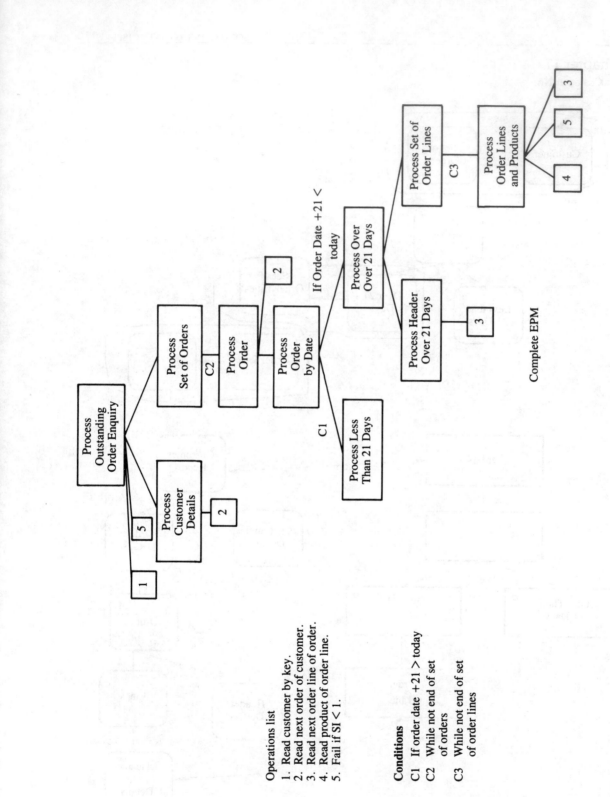

Operations list
1. Read customer by key.
2. Read next order of customer.
3. Read next order line of order.
4. Read product of order line.
5. Fail if SI < 1.

Conditions
C1 If order date +21 > today
C2 While not end of set
     of orders
C3 While not end of set
     of order lines

Complete EPM

Figure S.18

376

Bibliography and selected reading

The source for all information on SSADM version 4 is: CCTA (1990) *SSADM Version 4 Reference Manual*, NCC Blackwell, Manchester.

Other sources of information regarding techniques and philosophies used in SSADM are listed below.

Boehm, B.W. (1981) *Software Engineering Economics*, Prentice Hall, NJ.

Checkland, P. (1981) *Systems Thinking, Systems Practice*, John Wiley.

Codd, E.F. (1970) 'A relational model for large shared data banks', *CACM*, **13**, 6.

Codd, E.F. (1974) 'Recent investigations into relational database systems', *Proceedings IFIP Congress*.

Date, C.J. (1986) *An Introduction to Database Systems: Volume 1*, 4th edn, Addison-Wesley, Reading, Mass.

De Marco, T. (1980) *Structured Analysis and System Specification*, Prentice Hall International.

Fagin, R. (1983) *Acyclic Database Schemes (of Various Degrees): A Painless Introduction*, IBM Research Report RJ3800.

Fagin, R. *et al.* (1982) 'A simplified universal relation assumption and its properties', *ACM TODs*, **7**, 3.

Gane, C. and T. Sarson (1978) *Structured Systems Analysis: Tools and Techniques*, Prentice Hall.

Hares, J. (1993) *SSADM for the Advanced Practitioner*, John Wiley.

Jackson, M.A. (1975) *Principles of Program Design*, Academic Press, London.

Jackson, M.A. (1983) *System Development*, Prentice Hall International, London.

Loomis, M. (1983) *Data Management and File Processing*, Prentice Hall International.

Martin, J. and C. Finkelstein (1981) *Information Engineering*, Savant Research Studies.

Mumford, E. (1983) *Participative Systems Design*, Manchester Business School.

Yourdon, E. (1989) *Modern Structured Analysis*, Prentice Hall International.

Yourdon, E. and L. Constantine (1979) *Structured Design*, Prentice Hall.

Index